FORMWORK

a practical approach

ONE WEEK LOAN

Peter S.McAdam
PhD, MIEAust, CPEng
and
Geoffrey Lee
BSc, CEng, MICE, CDipAF

E & FN SPON
An Imprint of Chapman & Hall

London · Weinheim · New York · Tokyo · Melbourne · Madras

Published by E & FN Spon, an imprint of Chapman & Hall, 2–6 Boundary Row, London SE1 8HN, UK

Chapman & Hall, 2–6 Boundary Row, London SE1 8HN, UK

Chapman & Hall, GmbH, Pappelallee 3, 69469 Weinheim, Germany

Chapman & Hall USA, 115 Fifth Avenue, New York, NY 10003, USA

Chapman & Hall Japan, ITP-Japan, Kyowa Building, 3F, 2-2-1 Hirakawacho, Chiyoda-ku, Tokyo 102, Japan

Chapman & Hall Australia, 102 Dodds Street, South Melbourne, Victoria 3205, Australia

Chapman & Hall India, R. Seshadri, 32 Second Main Road, CIT East, Madras 600 035, India

First edition

© 1997 Peter S. McAdam

Printed in Great Britain by TJ International Ltd

ISBN 0 419 22820 9

A catalogue record for this book is available from the British Library

∞ Printed on permanent acid-free text paper, manufactured in accordance with ANSI/NISO Z39.48-1992 and ANSI/NISO Z39.48-1984 (Permanence of Paper).

TABLE OF CONTENTS

CHAPTER 1: AN OVERVIEW

1

General principles, quality, safety, construction (design and management) regulations, structural requirements, strength, stability, stiffness, sliding, overturning, uplift, sidesway; **Expected loading and construction effects**, concrete pressure, material loads, construction loading, impact, environmental loads; **The construction process**, load action, eccentricity, load limits and failure, failure mode, progressive collapse, **Economy**, learning curve, construction joints, care and maintenance, information from project documents.

CHAPTER 2: MATERIALS AND COMPONENTS

16

Material properties, strength, stiffness, impact resistance, durability, weight, accuracy, compatibility, insulation, staining, damage; **Framing materials**, solid timber manufactured sections, steel, aluminium; **Formface materials**, plywood - types and selection, solid timber, steel, GRP, GRC, foam plastic, particle board; **Release agents**, types, uses; **Fixings**, nails, screws, bolts: **Proprietary equipment**, devices and components.

CHAPTER 3: GROUND FORMS

26

Problems of ground forms; **Edge forms**, keyed, dowelled, reinforcement continuity, waterstops, cantilevered; **Footings**, isolated, pedestal forms, strip footings, eccentric footings, kicker forms.

CHAPTER 4: WALL FORMS

36

General, single faced forms, **Double faced forms**, loading, hoisting, climbing, horizontal waler forms, load paths, vertical waler forms, form type selection; **Wall tying systems**, he-bolts, coil ties, she-bolts, through ties, snap ties, bar ties, installation; **Wall form fabrication and selection**, bracing wall forms; **Construction details**; plywood fixing, frame fixing, stop ends, construction joints, kickers and kickerless construction, spandrel wall, corner details, form junctions, surface features, groove forms, recess forms, cast-in features, penetration forms, controlling the top of the pour, access platform, hoisting forms, pilaster piers, separate construction of pilasters, wall thickness variations, access panels; **Proprietary systems**, details, types for large forms; **Curved walls**, plywood radii, large radius walls, stripping, proprietary systems, use of straight sections; **Single faced forms**, overturning, anchorage, tie backs, sloping, corner tying.

CHAPTER 5: COLUMN FORMS — 86

General, concrete pressures, accuracy requirements, stripping problems; **Rectangular columns**, proprietary, column clamps, purpose-made column forms; **Conventional forms**, use of clamps, corner details, arris forms, erection, horizontal walers, other clamping systems, perimeter strapping; **Two-part column forms**, edge connections; **Hinged forms**, details, eccentic hinge; **Large rectangular forms**, internal ties; **Circular column forms**, types, stripping; **General form details**, kickers, kickerless construction, starter bars, base sealing, marking the top, bracing, hoisting, access platforms; **Gang forming**, principles, details; **Special column shapes**, examples.

CHAPTER 6: SOFFIT FORMS — 112

Design principles, stages of construction and loading, construction philosophy, causes of collapse; **Conventional soffit forms**, use of plywood, framing, joists, bearers, eccentricity, lapping; **Soffit form footings**, multi-storey, on blinding slab, footing details, footing stiffness; **Support systems**, props, frames, types and fittings, multi-storey frames, modular systems, details of erection; **Construction details**, junction to walls, edge forms, construction joints, top-step forms, penetrations, minor set-downs, sloping forms, horizontal tolerances, tolerance gap; **Stripping**, types, undisturbed supports, back propping, reshoring, removing components; **Aluminium beams**, use of; **Floor centres**, loading, twisting, end support, stripping; **Quick strip systems**, examples, procedures for use.

CHAPTER 7: BEAM FORMS — 152

General, loading, pressures, transmitting loads, a dangerous case, instability, load paths, stripping principles; **Beam form bases**, plywood deflection, wide beams; **Beam form sides**, floor centre soffits, joists to beam sides, separate support systems, plywood support, tolerance gap; **Deep beams**, ties for deep beams; **Junction** of beams of differing depth; **Edge beams**, instability, safety, built-in anchors; **Upstand beams**; **Proprietary** beam support systems; **Concrete encasing steel**, precautions, problems; **Multiple Tee beams**, principles, techniques; **Waffle slabs**.

CHAPTER 8: STAIR FORMS — 172

Principles, loads; **Details**, strutting, riser forms, connection to walls, improved riser forms, metal edge riser forms; **Precast stairs**, forms.

CHAPTER 9: PERMANENT FORMS — 176

Definition; **Materials**, timber, galvanized steel, concrete, GRC, fibre reinforced sheet, foam plastic, cardboard; **Examples**: Ground forms, Wall forms, precast concrete, Column forms, Soffit forms, void forms, fibrous plaster, cardboard tubes, precast concrete, Beam forms, Stair forms.

BIBLIOGRAPHY — 187

INDEX — 188

FOREWORD

Formwork costs are a significant part of the total price of a concrete structure. The value of materials and equipment used in formwork may represent 60% of the builder's on-site investment, and formwork is often the critical activity that controls the total time of construction of the project.

To optimise formwork costs and minimise its construction times, the contractor needs knowledge of the guiding principles of safe and efficient formwork construction, insight into the relative merits of the many available methods, and knowledge of practical details for its efficient construction. This book, which is intensely practical, addresses these areas. In many cases more than one solution is given for a particular construction problem. The text is clear and readable, and the multitude of diagrams show both principles and practice in great detail.

This book is aimed at the hands-on people of the construction industry, the carpenters, the foremen, the supervisors and the inspectors. As a result it is also a basic text for the building educators and their students, apprentices, building cadets and construction management trainees. It belongs in the work kit of every one of them.

K.J. Lyngcoln
Executive Engineer
Plywood Association of Australia

PREFACE

Amongst the many trades on a typical building site, the role and responsibilities of the formworker are unique. There are few restrictions placed on his choice of working techniques. In contrast, other trades are constrained by the most precise directions. For the structural steelwork all sizes, connections, fixings and painting are defined in detail. Reinforcement grades, sizes, positions, laps and tolerances are all predetermined. Joinery is exhaustively detailed, colour schemes are prescribed, and furnishings selected.

Compared to this, the formworker is almost permitted to be a free spirit. Most times, the only constraints are mandatory requirements on the concrete surface quality and accuracy, together with the builder's demands on cost and time. Outside this, he chooses his own formwork system, selects his materials and components, and devises the general arrangement and the details of construction. With this much freedom of choice it is not surprising that there has been so much innovation in formwork construction. The formwork industry has always been quick to adapt new products to its own purposes. Sadly, however, very few of these details have been recorded so others can learn. As a result, many of them have been reinvented, perhaps many times. One of my aims in writing this book has been to record this information, to show the 'tricks of the trade'. All of the techniques described have been seen on sites, and been seen to be practical answers to a problem. Some older techniques have also been included where I have thought they can still be useful. The use of fibrous plaster for permanent forms is an example of this.

With the other trades on the site, the responsibility for the success or failure of what they do can be said to be shared between them and the designers of the building. The formworker does not have that protection. The latitude of action that he has in all that he does, brings with it the full responsibility for the quality and safety of the formwork. The formworker must understand all the matters that affect quality and safety, and devise the formwork accordingly. To aid this understanding this book generally approaches formwork from this viewpoint, the function, the forces and the consequences. After this, each area deals with details of construction.

This book is not concerned with the structural design of formwork; that is another and separate topic. Few sizes of members or fixings are given. This book is intended as a practical guide to efficient formwork construction. It is confined to the 'why' and 'how' of formwork.

However, no book can cover every detail and every situation. The formworker must look at every particular case and be sure that it will produce the required quality with safety at all times. If he is not certain, he should seek expert advice. The causes of collapse are many and the things that initiate them can often be subtle. One of the messages within the book is the need for continuous vigilance in construction, at formwork inspection and during the concrete placement.

Further, not every type of formwork has been described. Constraints of size and cost have limited the range of the book. More advanced and exotic types of formwork, like self-climbing wall forms and table forms have been left for another volume. Also, there is only a limited amount of information given on scaffolding, guard rails, access ways and ladders. These are usually the subject of regulatory requirements. Their absence from the text or the illustrations must not be taken to mean that they are unnecessary; indeed they are vital. It is the formworker's responsibility to ensure that all these safety requirements are met.

In writing this book I have received help and encouragement from my friends at the Plywood Association of Australia and the Cement and Concrete Association of Australia. My sincere thanks to these people. Another major factor in the completion of the book has been the continuous encouragement I have received from my wife Wendy. She has endured this seemingly perpetual distraction with equanimity and forbearance.

Peter S. McAdam

CHAPTER 1:
AN OVERVIEW

A number of fundamental principles apply to all formwork. They relate to the use of appropriate materials and standards of workmanship, construction for ease of erection and stripping, and care and maintenance of the formwork so that the maximum number of re-uses can be achieved.

Even though formwork is usually only temporary construction, as much care must be taken with this temporary work as is taken with permanent work. Indeed, because it is a work area as well as a mould for the fluid concrete, the requirements are more stringent and the responsibilities of the formworker often much greater.

The skilled and experienced formworker has knowledge that could contribute much to the optimisation of building costs and speed of construction but is rarely involved in either. The decisions that control formwork efficiency are usually made by others. The formworker is usually presented with a fait accompli. The formwork must then be designed to suit the dictates of the building design, the restrictions of the construction documents and the builder's decisions.

Although this book is directed principally towards the 'how-to' of basic formwork construction, some of the matters that relate the shape of the concrete to formwork efficiency, reduced costs and enhanced speed of construction are covered in the relevant chapters.

Three general principles govern formwork design and construction:

QUALITY
SAFETY
ECONOMY.

These three matters are not separate and unrelated. Experienced formworkers know that it is a false economy to reduce quality. Further, if the formworker feels safe, this will lead to more production and thus reduced costs.

Throughout this book, even if they are not specifically mentioned, these three principles are fundamental to all the matters described. In this chapter their further discussion will relate 'Quality' to the quality of the concrete structure being produced, 'Safety' to both personal safety and formwork loading, and 'Economy' to the matters that affect the total effective cost of formwork and the contribution of this to the total cost of the concrete structure.

The activity of formwork construction, its concreting and subsequent stripping, can also have a significant loading effect on the permanent concrete structure being built. The design engineer for the permanent structure may place restrictions on the formworkers activities. The formworker must ensure that full **INFORMATION** has been supplied on these and any other requirements that will influence the materials, methods of use and quality of the formwork.

A general list of the range of matters that should be included in the information given to the formworker is the last topic covered in this chapter.

QUALITY

Quality, as it relates to the formed faces of the permanent concrete structure, refers to two aspects: the accuracy of the concrete shape and the quality of the surface. Surface quality refers to texture and can, where specified, include consistency of colour.

For accuracy in the concrete, the starting point is formwork which must remain acceptably accurate when loaded. The loads come from the fresh concrete and the work in placing and compacting that concrete. Stiff formwork, that has a limited and predictable deflection under load, is essential.

However, formwork deflections are not the only source of deviations in the lines and planes of concrete faces. The permitted deviations (tolerances) in the materials and components used, and the expected innaccuracies in workmanship, also contribute to the total deviations. For example, minor steps can occur in the concrete face at positions where sheets of plywood abutt. These can be caused by the permitted variations in the manufactured thickness of plywood of the same nominal thickness.

Face Step in the Concrete Surface

Codes of practice and the contract documents give maximum permitted values for these total deviations. By definition these are called **TOLERANCES**. In the design and construction of the formwork care must be taken to ensure that the anticipated total deviations are smaller than the tolerances by a significant margin.

To minimise deviations, all formwork must have adequate means of alignment and adjustment both at the construction joints and throughout the formwork. In the following chapters, on the various categories of forms, the means of achieving this will be shown. They include such devices as simple wedges, screw adjustments on supports, and camber adjustments.

Lack of stiffness in the formface can also affect surface quality of the concrete in two ways. Firstly, flexible form faces will often fluctuate in response to the vibrators used to compact the concrete. The release agent, applied to the formface to prevent adhesion, may become emulsified and react with the cement. The result will be unacceptable dark patches on the concrete.

Secondly, formwork that deflects excessively will usually leak at its joints, and this in turn can have two effects. In the first instance the grout and moisture loss will result in honey-combing and dark hydration staining of the concrete face. Beyond this, the escaping grout will fill gaps in the form structure and between the edges of the plywood sheets of the formface. By effectively locking up these gaps, this grout loss makes stripping of the formwork more difficult.

The formwork is not the only factor that influences the achievement of quality in the concrete surface. Other important matters are the details of the concrete shape, the quantity and disposition of the reinforcement, the quality of the concrete used, the efficiency of its placement and consolidation, and the final acts of stripping the forms and curing and protecting the concrete. The production of high quality concrete surfaces is a special and extensive area of study in itself. It is a minor part of this volume, which is directed towards the achievement of accurate, safe and efficient formwork for general construction work.

SAFETY

CONSTRUCTION (DESIGN AND MANAGEMENT) REGULATIONS.

In Europe the responsibilities of the designers, supervisors and operatives concerning worksite safety have been defined and formalised in recent years following the Temporary and Mobile Worksites Directive adopted by the European Union Council of Ministers in 1992. In Britain these became law by introduction of the Construction (Design and Management) Regulations implemented in 1995 and similar legislation is in force throughout the European Community. There are a number of subsidiary sets of regulations relating to workplace safety, use of equipment, personal protection and handling of loads etc., which are all relevant to formworking operations.

In essence the CDM regulations, as they are widely known, seek to ensure that all hazards associated with a construction operation are identified and that all risks are assessed and reduced as far as is practicable before the operation commences. This is done in two stages.

Firstly, a worksite policy and procedure concerning all aspects of safety planning is implemented. The client appoints a single individual or a company as Planning Supervisor. One of the responsibilities of the Planning Supervisor is the compilation of a Health and Safety Plan which incorporates safety procedures developed by the various organisations on the worksite. This Health and Safety Plan has a permanent role in determining both how the structure is built and how it should be maintained during its life span.

Secondly, at a more detailed level, a procedure and risk assessment are prepared for each individual operation, such as 'erection of formwork for wall pour in eastern bridge abutment' or 'concreting second storey floor slab'. In the early stages of a worksite these assessments can be quite time consuming but once a routine is established many of the operations are repetitive and previous assessments are updated for local or particular hazards such as the proximity of live electricity cables or handling problems caused by the limits of a crane's lifting capacity.

The formwork supervisor will liaise with the designer and safety supervisor concerning the design of the formwork, method of handling, erection and concreting, and will decide the construction sequence and safety precautions needed. Manufacturers' instructions will be needed for all proprietary equipment. Input from individual formworkers familiar with certain aspects of the work should be invited. Following the risk assessment the agreed procedure is then communicated

to the formwork team and the supervisor and operatives must ensure that the procedure is followed and that any necessary amendments are similarly evaluated before their adoption.

Safety has two major aspects: the personal safety of people, both the formworkers and the public, and the safety of the formwork structure.

Formwork assemblies, such as soffit forms for slabs, are a work area as well as a mould for the fresh concrete. When the formworkers have finished, steel fixers will place reinforcement and conduits and piping will be placed by electricians and plumbers.

For efficiency it is necessary that workers feel secure, and be secure, while working on this formwork. To this end, not only must the general form structure be safe, but the perimeter of the form must have effective guard rails with toe boards, access ladders and stairs, all secured in position. The formworkers should be equipped with safety helmets and boots. For special cases on high work, safety harnesses and security screens may be needed.

Working in Safety

The safety of the formwork structure starts with its design. This must include the evaluation of all the probable combinations of load at each of the stages of formwork use. In summary these are:

STAGE 1: before the placement of the concrete. This can be while it is being built, or while it is being relocated from another position, e.g. table forms.

STAGE 2: during the placement of the concrete.

STAGE 3: after the concrete placement when it continues to provide support until the concrete can support itself.

Example of Stage 1 Loading

Example of Stage 2 Loading

The formwork structure must not become unsafe at any time during any of these three stages. None of the components should be dislodged, become loose or unwind under the influence of the construction work. For example, the vibration from immersion vibrators may tend to cause the screw jacks on frames to wind down. This does not usually occur with well designed equipment. Similarly, vibration can loosen wedges; they should be nailed to prevent movement.

STRUCTURAL REQUIREMENTS.

There are two important aspects to structural safety: **STRENGTH** and **STABILITY**, and a further one needed so the form can produce an accurate structure: **STIFFNESS.**

STRENGTH simply means that the formwork structure must be able to safely carry the anticipated loads. This does not only mean being safe when all the loads are applied, (Figure 1.01) but also being safe when the structure is only partly loaded and when the load is progressively increasing. (Figure 1.02).

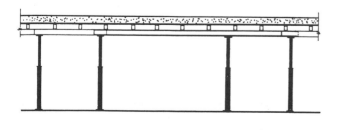

CONCRETE POUR COMPLETED

Figure 1.01 - FORM FULLY LOADED

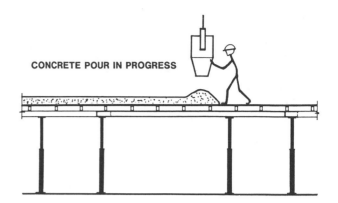

CONCRETE POUR IN PROGRESS

Figure 1.02 - LOADING PROGRESSING

The illustrations show a simplistic example which does not have any apparent problem in catering for the increasing loading. Other examples where this situation can be quite critical are given later in the book.

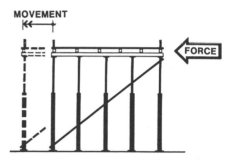

MOVEMENT

FORCE

Figure 1.03 - SLIDING OF THE FORM

For STABILITY four cases are examined: **Sliding, Overturning, Uplift,** and **Sidesway.** These problems are most often found in soffit form structures. Sliding is the movement of the total form sideways (Figure 1.03) This can occur by wind or water action. Sliding can also occur in a part of a formwork arrangement.

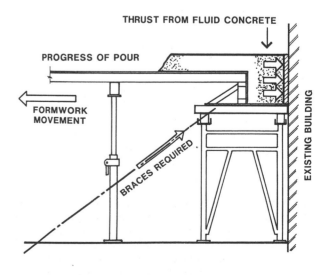

THRUST FROM FLUID CONCRETE

PROGRESS OF POUR

FORMWORK MOVEMENT

BRACES REQUIRED

EXISTING BUILDING

Figure 1.04 - BRACING AGAINST MOVEMENT

Figure 1.04 shows a soffit form where the edge beam is poured against the wall of the adjacent property. The pressure of the concrete against the wall causes that part of the formwork to slide with resulting misalignment of the inner beam face. A row of extra diagonal braces is needed to resist the sideways movement of the beam formwork and its supports.

Overturning of a soffit form is shown in Figure 1.05. Tall, narrow form assemblies with the lightly loaded formwork of Stage 1 are vulnerable to wind loading. Guy ropes are usually needed until the weight of the concrete improves the stability in Stage 2. (Figure 1.06) For very tall forms the guys may have to stay in place until the concrete has developed strength and it can provide the bracing effect.

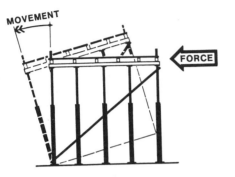

Figure 1.05 - FORMWORK OVERTURNING

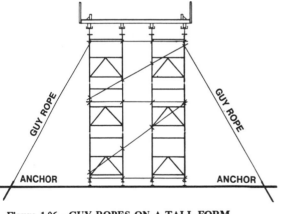

Figure 1.06 - GUY ROPES ON A TALL FORM

Figure 1.05 also shows the form uplifting on one side. Cases of uplift can occur, in a wide range of situations, to whole form assemblies and within form assemblies. A simple example is shown in Figure 1.07 where the overloading of one span of a two span continuous beam causes the unloaded end to lift up. If the prop is not secured it will fall over.

Figure 1.08 shows Sidesway. This is simply the result of inadequate bracing. Because horizontal forces can act from any direction, bracing must be provided in at least two directions at right angles to each other. Where permitted, connection to previously constructed permanent work can be a most efficient form of bracing. Otherwise the formwork assembly must be braced by effective framing within the formwork assembly.

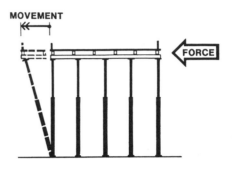

Figure 1.08 - SIDESWAY

Stiffness of the formwork is its ability to resist deformation under load. Design for stiffness means more than just catering for the deflection of the various members when fully loaded. It also includes consideration of progressive increases in deflection with increasing load.

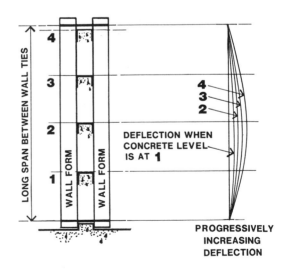

Figure 1.09 - PROGRESSIVE WALL FORM DEFLECTION

Figure 1.09 shows an example of a wall form where the wall ties have been specified to be a large distance apart vertically; perhaps even 4 metres.

The most important design aspect of this form is progressive deflection. As the pour progresses up the form, the bending effect on the form will increase with resultant increases in deflection. This will only be a minor problem while the full depth of the concrete remains fully fluid. The mass of fluid concrete will move the very small amount to remain in contact with the form face.

However, when the initial set of the lower concrete commences before completion of the pour, then any

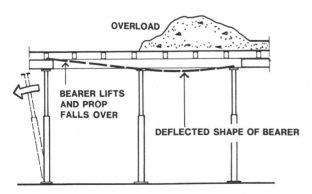

Figure 1.07 - UPLIFT OF THE END OF A BEARER

further progressive deflection of the form will move the formface a very short distance away from the stiffened concrete. Surface discolouration of the concrete will result.

Very stiff formwork framing is needed to minimise this effect. The deflection limitations for such cases must be much smaller than those permitted for wall forms where the wall ties can be closer together.

A similar situation has to be guarded against in tunnel forms. (Figure 1.10) As the concrete placement progresses the forms are subject to uplift and bending actions that tend to distort the shape. Stiffness is a paramount design criterion.

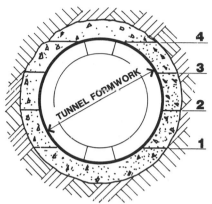

Figure 1.10 – TUNNEL LINING FORMWORK

Every part of the formwork assembly has a requirement for an appropriate level of stiffness. Footings, a critical part of soffit forms, also have stiffness requirements. These must be constructed to give a uniform, predictable and small movement under load. Differential movement as shown in Figure 1.11 is not acceptable.

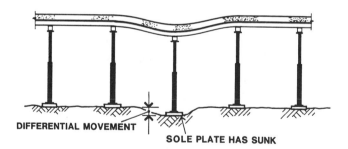

Figure 1.11 - DIFFERENTIAL SUPPORT MOVEMENT

Further, foundation soil movement over a period of time, known as creep, is totally unacceptable. This is not only results in more deformation of the form but also deformation of the concrete while it is setting. Such movement can make the concrete structure unserviceable. In that case, demolition is the only course of action. The problems of footings to soffit formwork are covered in much more detail in Chapter 6.

Expected Loading and Construction Effects.

This book is concerned with the practical aspects of devising and constructing formwork, and not with the structural calculations of formwork design. Nevertheless, it is important that the nature and sources of the many loads that act on the formwork be understood, so that a safe and workable formwork assembly can be built.

The total loads that act on the formwork should not be thought to be limited to the precise sum of the load of the documented concrete shape, the formwork, the building equipment and expected environmental effects (e.g. wind). In the real world of formwork, things do not always go to plan; concrete can be temporarily overpoured, the wind can exceed expectations and equipment use can cause bigger loads than anticipated.

Care must be taken when estimating loads, and reasonable and realistic overloads must be considered. Allowance must be made for the reasonably forseeable abuse of equipment as well as normal procedures.

1. Self weight

All formwork systems carry their own weight at all stages. Most reusable formwork is relatively light, such as that made from plywood and timber, but there are a few cases where the self weight can be significant. Examples of this are large structural steel forms for civil engineering projects and precast concrete permanent formwork units. (Refer to Chapter 9) These loads act in Stages 1 and 2 and for at least part of Stage 3.

2. Concrete loads on vertical forms.

Formwork for vertical concrete elements, such as columns and walls are subject to pressures on the formface from the fluid action of the fresh concrete. This is a Stage 2 load. The pressure of the fluid concrete on the vertical faces increases proportionally with the depth of concrete; the maximum pressure being at the bottom of the form.

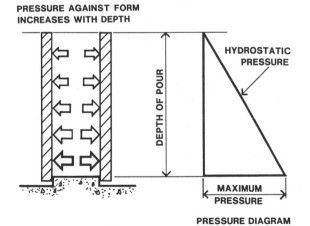

Figure 1.12 - FLUID CONCRETE PRESSURE

This maximum value for the pressure for the full depth fluid concrete is the hydrostatic pressure for concrete and usually occurs when the concrete is placed very quickly.

However, with slower pours the concrete at the bottom will gain its initial set and this maximum pressure reduces. The bottom concrete is no longer fluid. Figure 1.13 shows the pressure graph at four points in time during a continuous pour. As the pour progresses towards the top of the form, the lower concrete is getting older and its active pressure is reducing further.

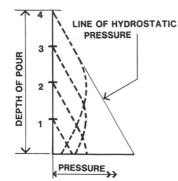

Figure 1.13 - CONCRETE SETTING REDUCES PRESSURE

If the pouring rate is very fast, then the concrete will have less time to set before the pour is completed, and the hydrostatic pressures on the forms will not be reduced very much. (Figure 1.14)

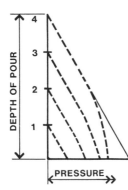

Figure 1.14 - PRESSURES WITH A FAST POUR

Conversely, a slow pour gives more time for setting of the concrete with resulting lower pressures. (Figure 1.15)

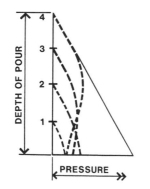

Figure 1.15 - PRESSURES WITH A SLOW POUR

Variations in the basic rate of setting of the concrete will also affect this maximum design pressure. A fast setting rate would give a lower design pressure, and a slow setting rate, which will result if the concrete temperature is low, will give a higher pressure.

The factors that have been found to control the design pressure are the concrete density, the pour height, the vertical rate of pour, the height of the discharge of the concrete, the concrete temperature, the cement type, the admixtures in the concrete and the plan dimensions of the concrete element being poured. The effect of these is summarised in Table 1.1, below.

TABLE 1.1 FACTORS CONTROLLING CONCRETE PRESSURE ON FORMS.

A. Concrete density	All pressures are proportional to the density of the concrete with normal compacted concrete having a density of 2400 kg/m³
B. Pour height	Higher forms have proportionately greater hydrostatic pressure.
C. Vertical pour rate	Faster pours lead to less pressure reduction effect from the initial set of the concrete.
D. Height of discharge	Where the concrete is discharged at a height greater than the top of the form there is an impacting effect from the drop which can increase the pressures near the top of the form.
E. Concrete temperature	Cement hydration, and therefore concrete setting, takes place more quickly at higher temperatures. The chemical process generates heat and this may be lost too quickly if the air temperature is very low, or too slowly to prevent cracking if the concrete pour is very thick.
F. Cement type	Cements blended with materials such as fly-ash or slag-ash set slowly and thus do not limit the pressure as much as early setting pure cements.
G. Admixtures	Both retarders and superflowing additives slow down cement hydration and concrete setting, prolonging the period of full hydrostatic pressure.
H. Plan dimensions	Where **BOTH** the width and the length of an element are less than 2 metres (e.g. columns and short walls) higher than normal pressures have been observed to occur. This is believed to be caused by the reflection of the vibrations from the formfaces. (Figure 1.16)

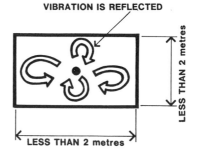

Figure 1.16 - PRESSURE IN CONFINED SPACES

The formworker should note that none of these pressure modifying factors are under his control. Therefore, the wisest and safest course of action is to assume that the concrete will be placed very quickly and build the forms to cater for the **full height hydrostatic pressure.**

It is noted above, that, for normal concrete placing techniques, the maximum pressure that can occur is full height hydrostatic pressure. However, where the placing technique is different then higher pressures can sometimes occur. There are at least two cases of this.

The first is full depth revibration. Here the concrete is poured full height without any vibration. The pour is done very quickly so that initial set of the concrete has not occured when the vibrators are inserted full depth into the concrete and then slowly withdrawn.

This can result in pressures that are up to one and one half times the full height hydrostatic pressure. This pressure increase is caused by the entrapment of the large energy input from the vibrators by the weight of the full height of fluid concrete above it.

The second case is the pumping of the concrete into the base of the formwork. (Figure 1.17) Here the pressure is far greater than the hydrostatic pressure expected from the same height of fluid concrete. The pressure increase comes from resistance to the rise in the concrete level as it is forced past the reinforcement and from friction between the rising concrete and the form faces.

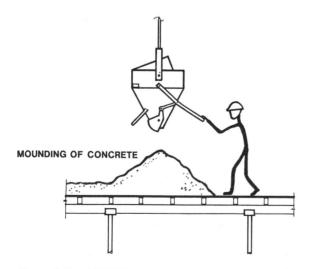

Figure 1.17 - PUMPING UP INTO THE FORMWORK

3. Concrete loads on horizontal forms.

The concrete load applies for all of Stage 2 and Stage 3. With the exception of very deep beams, which additionally have horizontal pressures similar to walls, the concrete loads on slab and beam formwork relate only to the depth of concrete placed on the form.

This depth of concrete will often temporarily exceed the intended slab depth. Malfunction of the discharge gate to a concrete skip can easily cause mounding. (Figure 1.18) Even more common is the case where the concrete pump operator does not respond to signals to stop the pour.

The load from this excess concrete is only a short term load, usually less than five minutes. This is significant for timber framed forms. The duration of the load is an important factor in timber design. Checking the structural adequacy of framing members for this short duration load from the mounded concrete can be done against a small increase in permissible stress.

The mounding of concrete from concrete boom pumps and skips may even be deliberate. If the boom will not reach the far corner of the slab then the concretor may mound the concrete as near as possible to the corner. From there it is shovelled into position. (Fig 1.19)

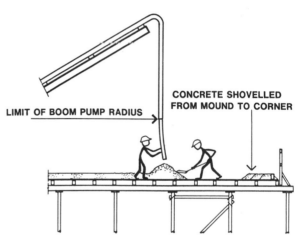

Figure 1.19 - DELIBERATE MOUNDING OF CONCRETE

4. Material loads on horizontal forms.

Horizontal formwork for slabs and beams is often the work and storage area for its own progressing construction. Completed parts of the form will be used to stack materials, plywood and timber or formwork components, for the construction of the next section of the formwork. These are generally Stage 1 loads.

Material Stacked on the Formwork

Figure 1.18 - ACCIDENTAL MOUNDING OF CONCRETE

Materials Stacked on Site

When the surface of a form is finished, material for the next trade, the reinforcement fixing, is often stacked on it. If the support structure is not complete then care must be taken in selecting the position for placing the reinforcement bundle. Figure 1.20 shows a section through such a structure.

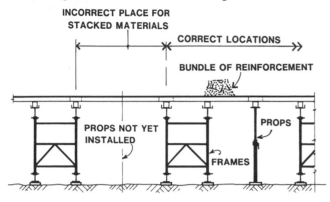

Figure 1.20 - LIMITATION ON PLACING MATERIALS

When completed the form will have a row of props and bearers between each pair of lines of frames. Steel fixing often starts before these rows of props are in position. Obviously the heavy bundles of reinforcement must not be placed over the long unpropped spans. In such cases the permitted locations for stacked material must be clearly marked on the formwork surface.

Another dangerous loading problem can come from stockpiling materials for later tradework in multi-storey buildings. (Figure 1.21)

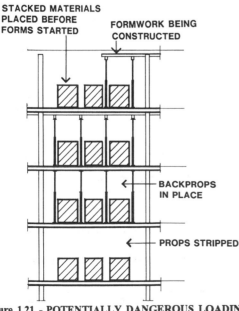

Figure 1.21 - POTENTIALLY DANGEROUS LOADING

To avoid difficult lifting problems, materials, such as bricks or blocks for partitions, are often stacked on newly poured top slabs before the next level of formwork is constructed.

Most of the extra loads are carried down through the formwork system to lower levels of the structure and are therefore Stage 3 loads. The formwork supports are carrying the imposed loads of this material as well as accumulated dead load of the floors, which are equal to at least two slabs. The load capacity of the formwork supports must be at least equal to this greater loading.

These additional loads have the potential to overload most normal multi-storey concrete structures. Material placement of this type should not occur without careful consideration and the permission of the design engineer for the building structure.

5. Formworkers and tools.

The loads of workers and their equipment can occur at all loading stages. When the workers are on the formwork and are building it, it is Stage 1, when they are placing the concrete it is Stage 2, and when they are on a newly poured slab and are starting to build the form for the next level of the multi-storey building it can be Stage 1 and Stage 3. (Figure 1.22)

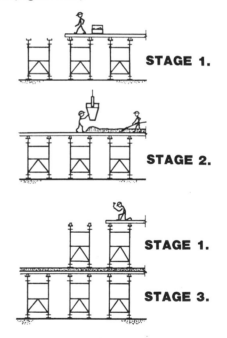

Figure 1.22 - CONSTRUCTION LOADING

6. The work process - Impact

On most projects there is crane handling of bundles of material, concrete skips and large formwork components. Although the probability may be small, there is always the possibility of collision between these crane loads and parts of the formwork.

The crane loads can horizontally strike the tops of wall or columns, and the edges of soffit forms. Vertical impact loads can shatter the formface of soffit forms. The effect of these impact loads must be considered when arranging the formwork system. These three situations are illustrated in Figure 1.23.

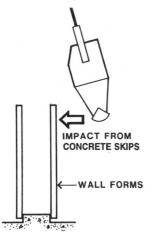

Figure 1.23/1 - IMPACT WITH WALL FORMS

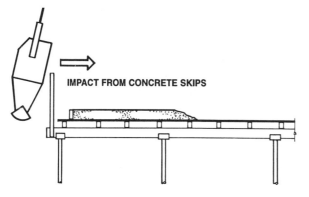

Figure 1.23/2 - HORIZONTAL IMPACT ON SOFFIT FORMS

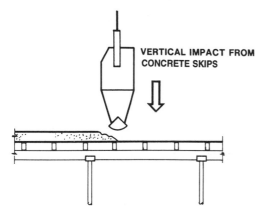

Figure 1.23/3 - VERTICAL IMPACT ON SOFFIT FORMS

Only where the formwork is a very heavy structure, or where the weights being hoisted are quite small, can the formwork be assumed to be able to withstand the impact without significant damage.

For normal formwork, columns, walls and soffit forms, major damage and failure will most probably result from the impact from heavy weights. If there is a pour in progress in the damaged formwork, then the pour must be terminated and rectification work undertaken.

The most important aspect of this failure is that the site personnel and the general public must not be endangered as the formwork fails under impact. It is essential to ensure that the form, even though badly distorted and damaged, does not generate dangerous debris. This topic

of the **'Failure Mode'** is discussed later in this chapter.

Not all impact loads are major forces with such a dramatic effect; many of them are quite minor. These can be materials, e.g. planks and timber, being stacked against the form, or the loads occasioned when tools and equipment are manhandled into place. In short, the normal activities of building. The bracing that must be provided for wind and other horizontal loads must also cater for these minor construction loads.

7. The work process - The sequence of the pour.

Unless appropriate provisions are made, the direction and sequence of the pour can often determine whether the pour will be a success or a failure. At the least, if control is not exercised, there might be excessive deformation of the form. One example of this is a circular wall form. If the pour proceeds uniformly up the form for the full perimeter then the deformation will be minimal, usually insignificant.

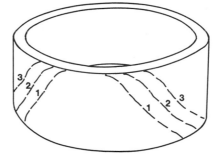

Figure 1.24 - UNEVEN POUR OF CIRCULAR WALL

However, if the wall pour is brought up to near full height at one point before proceeding to be poured at other parts of the wall, then the wall form may well distort out of the circular shape. This incorrect pouring sequence is shown in Figure 1.24. To resist the distortion caused by this procedure, adequate and extensive bracing and ties would be needed to maintain the circular shape.

Of a more serious and dangerous nature are cases where structural stability is involved. Figure 1.25 shows formwork for a slab that cantilevers beyond its base of support.

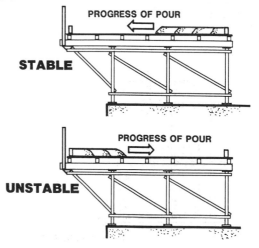

Figure 1.25 - IMPORTANCE OF DIRECTION OF POUR

If the pour starts from the right-hand side then the situation will be **STABLE**. The load of the fresh concrete will increase the stability of the formwork assembly. When the last concrete is placed out on the cantilever its overturning effect will be resisted by the weight of the concrete placed earlier on the right hand end.

But if the pour commences at the left-hand side then the weight of the empty form of other end will not be able to resist the overturning. The form will be **UNSTABLE** and failure will result. Safety, in this case, can only be assured if the right hand end is adequately anchored down.

The formworker must always remember that he will rarely have any control over the concrete pour. He should always assume that the pour will procede in the most potentially hazardous way and construct the form accordingly. In this case anchors, as noted above, or counterweights are needed on the right hand side of the formwork.

8. Environmental loads - Wind

All formwork must be braced to cater for wind loads. The extent of these loads will vary according to the local climate and the degree of exposure of the formwork. Figure 1.26 shows the contrasts that can occur in the exposure of formwork to the wind.

FORMS AT 'A' HAVE HIGHER WIND LOADS THAN FORMS AT 'B'

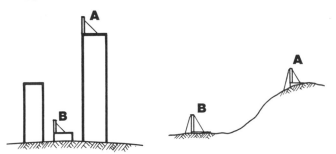

Figure 1.26 - DIFFERENT EXPOSURES TO WIND FORCES

Forms on the top of tall buildings or located on high positions will have less shielding than those on level areas and located between existing buildings. The wind pressure on formwork in high exposed positions, **'A'**, can be up to three times that acting on sheltered formwork, **'B'**.

9. Environmental loads - Water.

Both rain, and the runoff that can result from it, can have a detrimental effect on the formwork structure. In the first instance rain can soften the soil of the foundation, and the formwork footings may sink causing misalignment and even instability of the formwork structure.

Secondly, if the quantity of water is considerable, there is the potential for the water flow to scour out the soil at the formwork foundations. Where this possibility exists, diversion drains and protective levee banks should be constructed. This situation can occur during any of the three stages of loading.

Load Action

The next, and equally important, matter related to loads on the formwork structure is 'How do they act on the formwork support structure?' and the answer is **'Eccentrically'**. Figure 1.27 shows a simple example of eccentric action, that is, off-centre loading. Because the

load is not directly in line with the prop there are two reactions to the load. That is, two types of resistance to the load. The first is compression in the prop and the second is bending.

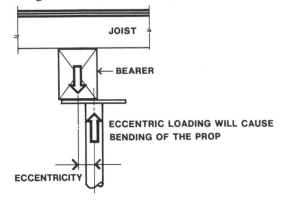

Figure 1.27 - ECCENTRIC LOADING OF A PROP

Even if great care is taken to load the prop axially there will always be some eccentricity. This may occur because the timber bearer is not quite square and it will tend to sit on one corner. (Figure 1.28)

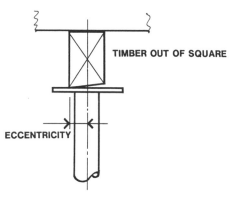

Figure 1.28 - ECCENTRICITY: BEARER OUT-OF-SQUARE

In other cases the prop may not be quite straight. (Figure 1.29) The tubes that it is made of were never perfectly straight, the welding in its manufacture can cause small distortions and the manner of its use, and abuse, can cause small bends in the prop.

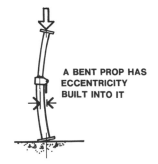

Figure 1.29 - ECCENTRICITY: BENT SUPPORTS

Care should be taken to minimise all eccentricities but there are many causes: out-of-plumb props, Figure 1.30, uneven concrete slabs, Figure 1.31, hard spots under the footing system, Figure 1.32, or simply poor construction methods, Figure 1.33.

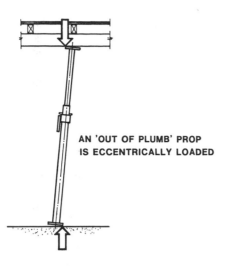

Figure 1.30 - ECCENTRICITY: OUT-OF-PLUMB PROPS

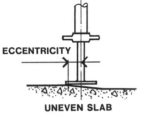

Figure 1.31 - ECCENTRICITY: UNEVEN CONCRETE SLAB

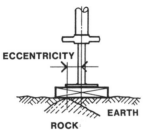

Figure 1.32 - ECCENTRICITY: UNEVEN FOUNDATION

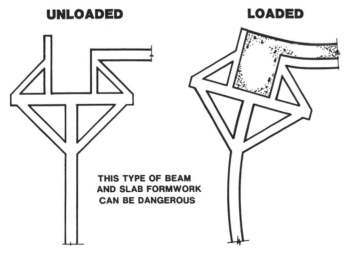

Fig 1.33 - ECCENTRICITY: DANGEROUS CONSTRUCTION

The last example, Figure 1.33, is especially dangerous. While the eccentric effects of the examples shown in Figures 1.27 to 1.32 inclusive can be minimised by care in construction and component selection, the method shown in Figure 1.33 is inherently hazardous and should never be used.

This topic of eccentric load action will be discussed again, and in more detail, in later chapters.

Load Limits and Failure

The statement has been made several times that overloads can and do occur. They can be overpouring concrete, high winds or simply the result of human error. Good construction practice calls for the formworker to build formwork that can withstand reasonably forseeable amounts of overload and levels of abuse of equipment and construction procedures.

The discussion on IMPACT loading showed that loads greater than the strength of the formwork might well occur. Consideration must be given to controlling the way in which the form will fail, the **FAILURE MODE**.

This approach to designing forms to have a predetermined FAILURE MODE must not be confined to those that might be subject to impact; it must be apply to all formwork. Formwork failure can occur from many causes. In devising a formwork system, thought must be given as to how it will collapse if any of the types of loads become large enough for failure to result.

In general, there are two types of failure: gradual and precipitate. A simple example of three storey formwork frames carrying a slab soffit form can illustrate the difference.

It is recommended practice that the successive stories of frames be connected to each other, at all legs, with the special pins or bolts and nuts. The reason for this becomes obvious when the failure of these frame towers is observed during product testing.

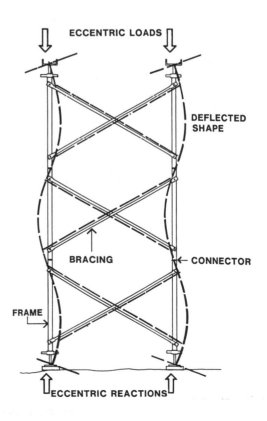

Figure 1.34 - GRADUAL BUCKLING FAILURE

As the frames fail, the buckling action tends to pull these connections apart. (Figure 1.34) If the connecting pins are in place, the failure will be 'serpentine' and relatively gradual. Without the pins the frames, whose legs were sleeved one into the other, will move apart as the frame buckling commences. Figure 1.35 illustrates this. The failure will be precipitate; a rapid and dangerous collapse.

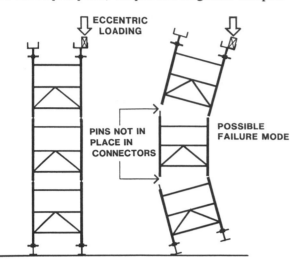

Figure 1.35 - RAPID FAILURE

The rapid action of a precipitate collapse usually causes a 'shock' load transfer to the other parts of the formwork structure immediately adjacent to the area of first failure. **PROGRESSIVE COLLAPSE** is often initiated. This is where the additional load, on the members adjacent to the first one that collapsed, is sufficient to cause it to fail. In turn, this member collapses under the excessive load and the area of failure rapidly spreads across the formwork assembly.

Result of Progressive Formwork Collapse

Two examples of progressive collapse can be given. Figure 1.36 shows a soffit form assembly.

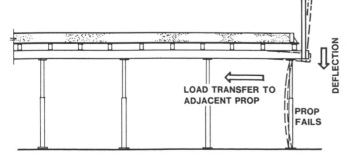

Figure 1.36 - SOFFIT FORM: PROGRESSIVE FAILURE

If the prop at the outside fails, and this could be from impact or excessive eccentricity of loading, then the formwork framing of joists and bearers will transfer the load to the adjacent props. At the same time, the deflection of the formwork beares or joists, which have become a cantilever, will twist the head of the prop and it is also liable to fail. If this happens it will be the start of a progressive failure.

The other example is more common, the progressive failure of wall ties. This usually starts with the failure of one tie. Wall forms are rigid structures and the load that was carried on the failed tie is rapidly transferred to the adjacent ties. The shock transfer of the load to these ties often results in their failure. The sequence of failure may proceed along the whole length of the wall form. Wall tie failure is most common where deep revibration of the concrete is done.

In summary, the formwork constructor must think beyond the case of the formwork being able to doing only the calculated structural task. Overloads do occur, and occasionally they can cause failure. Often this failure is initiated by minor defects, or minor mistakes in construction.

To ensure that any failure that occurs causes only the minimum damage and hazards, the formwork should be arranged with the failure mode in mind; a failsafe structure, one that does not readily tend to a progressive collapse if overloaded.

ECONOMY

In building construction it is usual for the structural frame, even for buildings of only one storey, to be the most significant cost component and a dominant and critical factor in the time of construction. When seeking the **"best"** method of construction the four components of the cost of a concrete structure must be considered. These are:

> **CONCRETE,**
> **REINFORCEMENT,**
> **FORMWORK and**
> **TIME.**

The details of the concrete shape and the reinforcement design are rarely under the control of the formworker, but the formwork design and the time effects of that design are. Time effects are more than just the total man-hours of fabrication, erection and stripping of the forms. They also include the cost effects of the total number of days that the formwork activity adds to the whole building program.

This latter aspect includes, in its often considerable costs, the expenses of site administration, plant hire and the cost effect on the cash flow of financing the building.

Where formwork is a repetitive activity, such as in the construction of a multi-storey building, a small reduction in the repetitive cycle time can result in large overall savings.

For a single use of a formwork system, there can be four components in its total cost. The first is the cost of fabrication of the formwork, its materials and labour, The second cost involves erection and completion of the formwork, labour costs and some hoisting. After the concrete pour the next activity is stripping, repairing and cleaning the formwork. This also involves labour costs and hoisting.

If the formwork is stripped in large units, such as wall forms, then there will be the fourth cost. This involves dismantling these units to recover materials and components for later reuse. The more often we can re-use materials, the less their unit cost of use in each form becomes.

If the same formwork assembly is to be repetitively used there can be large savings. The fabrication costs, increased slightly by some maintenance work, are spread over the number of uses. Erection costs and stripping costs will reduce with each successive re-use due to the 'Learning Curve' effect. As Figure 1.37 shows, at each repetition of the task, the time spent lessens, especially at the early repeats.

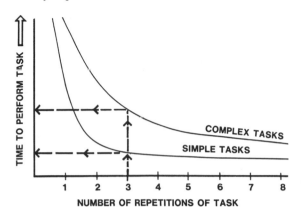

Figure 1.37 - A LEARNING CURVE

As the graph indicates the improvement in times with each repetition is more pronounced with simple tasks than complex ones.

The savings from repeated uses can be considerable and, in cases where there is a large number of repeat uses, consideration should be given to using a more sophisticated formwork system. Even though the first cost of fabrication may be high, the better system may require less maintenance, and may also afford faster construction cycles and so achieve a net cost saving.

The circumstances that can make the use of a special formwork system viable are:

(1) If a large number of re-uses is planned
(2) If the concrete shape is too complex for satisfactory site fabrication of the forms
(3) If the specified spacing of the wall ties is large and requires long span wall forms
(4) If concrete placing techniques or the type of concrete used are likely to result in very high pressures
(5) If there are difficulties with hoisting and the formwork unit has to incorporate its own means of movement
(6) If the tolerances called for are unusually stringent
(7) If the economics of the contract are such that the value of time saved more than offsets the increased cost of the formwork.

Whether it is used once or many times, the formwork system must have ease of fabrication, erection, adjustment and stripping. The size of each unit must not exceed the available handling capacity, crane or manhandling. A good, simple, rugged formwork system with the minimum of parts, components and fixings usually gives the most economical arrangement.

For formwork which is to be fabricated for use in large units, the designing of the formwork arrangement starts with the planning of the sequence and parts of the pour. Usually this involves deciding where the construction joints are to be located.

Rarely does the formworker have the authority to determine the position of the construction joints. This is the responsibility of the design engineer for the structure. When planning the work sequence the formworker must confer with this engineer.

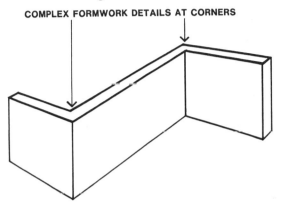

COMPLEX FORMWORK DETAILS AT CORNERS

Figure 1.38 - EXAMPLE OF WALL TO BE BUILT

Figure 1.38 shows a wall to be constructed. If no construction joints are permitted, it must be poured in one piece. The various parts of the formwork will have to be fabricated to suit, and there will be complex formwork construction details at the corners.

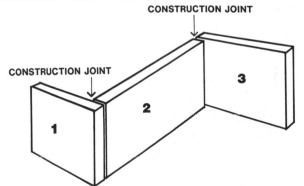

CONSTRUCTION JOINT

CONSTRUCTION JOINT

Figure 1.39 - CONSTRUCTION JOINT LOCATIONS

When construction joints are permitted at the corners, the three walls can be constructed separately. (Figure 1.39)

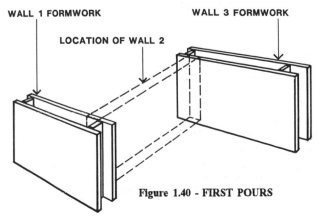

WALL 1 FORMWORK   WALL 3 FORMWORK

LOCATION OF WALL 2

Figure 1.40 - FIRST POURS

These construction joint locations permit the use of existing forms which may be much longer than the concrete walls. Further, WALLS 1 and 3 can be poured at the same time. (Figure 1.40)

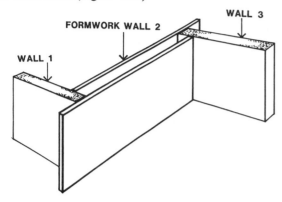

Figure 1.41 - POUR OF WALL 3

After stripping these forms, WALL 2 can be formed and poured. (Figure 1.41)

Although this sequence may take longer than the time taken to form and pour this complex wall in one stage, it has the advantage of using less formwork, and incurs less formwork material costs. As the diagrams imply, forms from other work can often be used and this further reduces costs.

In almost every project, the formworker must carefully analyse how the total structure can broken down into separate elements to achieve the most economical formwork construction procedure.

Looking now at a suspended structure. Figure 1.42 shows a simple example of a suspended floor supported on a wall and columns with a ground slab and simple footings.

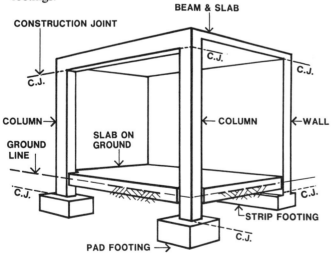

Figure 1.42 - EXAMPLE OF SUSPENDED STRUCTURE

Construction joints are shown at the tops of footings, walls and columns. One important decision will concern the time to pour the slab on the ground. Obviously, it must be placed after the columns and wall; but should the suspended slab be formed and poured before it?

If the suspended slab precedes the ground slab then other trades can work on this upper area while the ground slab is being done. However, it will be easier, and cheaper, to seat the forms to the suspended slab on a concrete

surface rather than install soleplates on the rough ground.

Clearly, if the overall time of construction is not important, then the ground slab should be built before the suspended slab. The ground slab will provide an accurate rigid base for the formwork supports. If they are to stand on the ground, then there will have to be adequate soleplates provided. However, if other construction considerations apply then the more costly formwork decision may be the best for the overall project.

Care and Maintenance of Formwork

Formwork is usually the most costly part of the concrete construction process. To minimise its cost, do not treat it as an expendable item. It is plant, even though it often has only a short life. Fundamental to getting the longest effective life out of it is proper care and appropriate maintenance.

Details of the factors controlling efficient formwork removal (stripping) are covered in the relevant chapters. For maximum formwork life, stripping must be done with care. Of equal importance are the actions that immediately follow stripping: handling, storage and maintenance.

After stripping, either as large form units or as individual components, the formwork should be moved away from the work area. If left there, it can be damaged and will be an obstruction to other trades. It should be stored where it will not deteriorate, preferably out of the weather. To avoid surface damage, large wall forms should be stood in racks, facing away from direct sunlight. (refer to Figure 4.115)

All formwork assemblies and components become encrusted with some concrete. There with be, at the least, dust on the formfaces, hardened slurry on the edge of plywood sheets and concrete in the mechanisms of equipment. All this must be removed before it hardens further.

Mechanisms and locking devices of equipment must be cleaned and oiled. Fixings must be checked and repaired on large formwork assemblies. Hardened slurry on the edges of plywood sheets should be removed, the surface cleaned and given one coat of release agent. Care must be taken not to damage the presealed surface of the plywood. Only a soft brush is suitable for dust removal. If larger particles of concrete have adhered then don't remove them with a metal tool. A softwood timber wedge is best; use it carefully so the surface is not broken.

In summary, care of formwork will prolong its useful life and, thereby, greatly reduce formwork costs.

INFORMATION

Many formworkers believe that the only interest that the design engineer takes in the formwork process is to place restrictions on how early the formwork can be stripped. They know that to remove the forms before the concrete has developed the appropriate strength is to court failure or, at the least, excessive deflection. However, the effects of the formworkers actions on the structure are not just limited to this obvious case.

For example, the loads from his stacks of materials and equipment often pass down through previously constructed work, and this accumulated load can often exceed the working load of parts of the building. More commonly, the locations of the construction joints may be compatible with the structural action of the permanent concrete structure or they may cause dangerous secondary effects.

On the subject of tolerances and surface finishes, full information must be provided. It is essential that the specification is clear and unambiguous.

In short, there are many constraints on the formworker's decision making and it is vital that the project documentation covers all matters that affect them. This includes a clear definition of the concrete shape and its surfaces, and limitations on formwork activities that can affect the strength, stability and serviceability of the permanent structure.

On the topic of the concrete shape and its finishes this would include:

(1) All dimensions, tolerances, and lines and faces for the assessment of tolerances

(2) Details relating to surface finish, colour control, surface treatment (e.g. sandblasting), and rules on repairs

(3) Requirement on the protection of finished concrete surfaces from damage.

On the topic of limitations on the formworker's activities this would include:

(1) Minimum and maximum stripping times and stripping procedures

(2) Procedures for the determination of the location of construction joints

(3) Precautions to be observed to avoid any detrimental effects of post-tensioning procedures on the formwork

(4) Any limitations on the extent of stacked materials either on the formwork or previously poured and stripped slabs

(5) Minimum and maximum requirements, procedures and precautions to be adopted in the forming, support and stripping of floors in multi-storey construction

(6) Any restrictions on the use of the previously poured permanent structure for bracing of the formwork

(7) Procedures and requirements for propping composite construction

(8) Guidelines on the sequence and method of concrete placement if this is critical to the structure

The lists given above are not exhaustive. Each structure may have its own unique features. If any of them affect the formworker's actions and decisions then that information must be available.

****

This chapter has only been an overview of the broad field of formwork for concrete buildings. Many important topics have been addressed, but none exhaustively. In the following chapters further information is given on the principles and practices of the main areas of building formwork. Even this is not a total statement. Formwork is an ever evolving and innovative industry. This book aims to cover the basics of 'why' and 'how'.

CHAPTER 2:
MATERIALS & COMPONENTS

Like the completed formwork, the materials of its construction must play their part in providing an accurate mould for the fluid concrete, carrying its weight and that of the workers and their equipment, and simultaneously resisting the forces from construction activity.

In the selection of materials for formwork, the three general principles of **QUALITY, SAFETY** and **ECONOMY** must be paramount. Material quality can ensure safety, and significantly contribute to the achievement of economy. Formwork failure can result in loss of life, and always causes catastrophic financial loss.

A Formwork Failure

Some general guidelines can be given for formface and framing materials, and for the associated components. These can be covered under the headings of **Strength, Stiffness, Impact Resistance, Durability, Weight, Accuracy, Compatibility and Insulation.**

1. Strength.

The material strength must be adequate to resist the forces anticipated. This is not only a structural design requirement, but also an essential safety aspect.

2. Stiffness.

The structural movement under load must be small and predictable. These deformations and deflections can be a significant part of the total deviations in the formed concrete surface. When the formwork designer is planning the formwork system, decisions must be made on the total deviation that will be acceptable, and to what extent workmanship errors and structural deformation will each contribute to this. To ensure that the total deviations do not exceed the tolerances, the material stiffness and the workmanship accuracy must be consistent.

3. Impact Resistance.

The possibility of major formwork damage occuring due to impact loads was briefly described in Chapter 1. As discussed, the forms must be built to ensure that the damaged form, although unserviceable, does not generate falling debris. It follows, that the way in which the formwork material fails, will determine this. To comply with this important safety aspect, materials exhibiting ductile failure are far superior to those that fail in a precipitate and brittle manner.

4. Durability.

In the interests of economy, and the achievement of a quality concrete product at each re-use of the formwork, its materials must be durable. Formwork is almost always built and used out in the open. Between re-uses, its materials and components are commonly stored out in the weather. Ideally, framing, components and formface materials should be resistant to the ravages of the environment. They should have a slow rate of deterioration under the effects of sun, wind and rain. Their resistance to deterioration can be enhanced by proper care and maintenance. Material durability is not only important for the achievement of good quality concrete surface finishes, but also to ensuring that formwork structures are always safe.

5. Weight.

In the assembly of formwork, most individual members and components are moved into position by hand. This occurs even when the completed formwork assembly is so heavy that it can only be moved and positioned by crane. Ideally, for efficiency and economy, framing members, formwork components and formface materials, should be sized such that their weight is within the lifting ability of one formworker. If the weight exceeds that which can be carried by two personnel, crane handling is called for. The next level of formwork weight restrictions is set by the lifting limitations of the on-site crane.

6. Accuracy.

For economy, it should be possible to assemble formwork with the minimum of fitting and cutting of materials. Consistency of size of materials, plywood sheets and framing members, is important to this aim. The accuracy of plywood sheets and the 'sizing' of timbers for consistent dimensions are discussed later in this chapter.

7. Compatibility.

The materials of the formwork must not be incompatible with either the fluid concrete or the hardened concrete. At the formface the constituents of the form materials must not react with the hydrating cement of the concrete. For example, some timbers contain wood sugars that break down the cement. After the concrete hardens some timbers, such as eucalypts, can severely stain the concrete. When water runs over this timber and onto the concrete, dark brown stains usually result.

8. Insulation.

Extremes of heat and cold present problems in the choice of form materials and their protection. The rate of setting of concrete and subsequent strength gain is slowed by low temperatures, and if the water in the mix becomes frozen, the formation of ice will destroy the chemical bonding within the concrete matrix. In situations where concrete has to be placed at low temperatures, aggregate storage bins and mixing water can be heated to produce warm concrete that will not cool during the initial setting period, while its own internal heat builds up.

Subsequently, newly poured slabs and wall tops can be covered with insulating blankets, but the soffit and the wall forms must not allow freezing air temperatures to penetrate the concrete surface behind. Plywood formfaces provide quite good insulation, provided they are free of ice when the concrete is first poured, but steel forms will rapidly conduct heat from the concrete. Steel frame panels with plywood faces offer poor insulation around the perimeters. The concrete around steel tie positions in any type of form will gain strength more slowly than the bulk of the pour.

Insulated formwork panels are available and lagging of steel forms can help but the lagging is susceptible to damage and it is very difficult to adequately insulate the corners of the form where heat can be lost in two directions. Columns and walls are best protected by movable enclosures with heating.

At the other extreme the formworker may have to allow for placement of cooling water pipes in thick concrete pours where heat loss is slow. Forms exposed to strong sunlight can also become extremely hot and cooling by water sprays or covering with wetted sheeting is often necessary.

For formface materials, two further general considerations can be added. Firstly, they should exhibit abrasion resistance, especially for soffit forms. Formworkers walk on them, and reinforcement and other formwork components are stacked on them.

For all forms, the placing of the fluid concrete, particularly with crushed rock aggregate, can cause some abrasive damage to the formface at first use and each re-use. When the concrete has hardened, the forms are stripped, and this can contribute to surface damage. Abrasion will occur if the forms are permitted to slide on the concrete face.

Secondly, moisture absorption at the formface must be minimised. Moisture loss from the concrete into the formface causes hydration staining of the concrete, with severely darkened surface patches. For high quality concrete surfaces, where colour control is specified, this is totally unacceptable. For any concrete surface, hydration staining means poor cement hydration, weak concrete and low surface durability.

Hydration Staining

In summary, it can be said that for economy of construction, better, more consistent and accurate materials lead to faster construction and a longer material life.

Further, formwork is most often built from materials that are held in stock. It is economic to limit the range of sizes of held in stock. This may lead to the use of some larger sizes than those dictated by minimum strength requirements. But this not a cost problem, as all the materials are recoverable.

However, high quality materials and components, even if very durable, cannot be re-used indefinitely. With each use there is some deterioration. For this reason, their quality and suitability for the intended task must be carefully re-evaluated at every use.

Plywood surfaces must be inspected for imperfections, edges assessed for delamination and damage, and the stress grade checked. The rough handling that can occur with stripping can cause splits in timber framing members. Components may have rusted excessively in storage and their locking mechanisms may no longer operate.

Components Stacked on the Formwork

Damaged Formwork Materials

No complete list of matters to be checked could be given; formwork materials and techniques are too diverse. In summary, at each use, every aspect of formwork materials and components must assessed for their suitability for the surface finish, structural and operational requirements.

FRAMING MATERIALS.

Almost all the structural materials used in general construction work are used in formwork construction; even reinforced concrete has a place in permanent formwork. Re-useable forms are mostly made of timber, steel or aluminium. Timber encompasses solid timber sections, general manufacture products such as plywood, laminated veneer lumber, and a range of special manufactured sections. Some of these special sections are shown in Figure 2.01. All those shown are a combination of plywood webs and solid timber flanges.

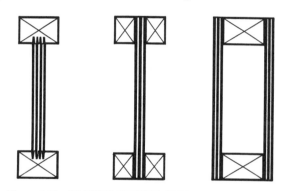

Figure 2.01 - MANUFACTURED SECTIONS FOR FRAMING

The following is an overview of materials used in the framing of re-useable formwork. Materials for permanent formwork are covered in Chapter 9. Further information on framing techniques used in the various types of formwork, is given in the relevant chapters.

1. Timber - Solid Sawn Sections.

Solid rectangular timber sections for formwork should be light, strong, stiff, durable and non-staining. The strength requirements include good resistance to lateral grain crushing, as well as consistent flexural and shear strength. They should be available in long straight lengths, with only a small amount of knots, splits and shakes. Usually formwork timbers are at least partially seasoned when milled. The further seasoning of the member should not result in warping or excessive shrinkage.

Species of Douglas Fir of North American origin have been found best to meet these criteria, but species of pine are also suitable. Where timber sections are to be used in support to the formface they should be bought 'stress-graded' to ensure that the appropriate design strength requirement is fulfilled. However, there is no advantage in over-specifying timber, since the formworker is seldom concerned with long term structural performance. 'General structural' grades to strength classification SC3 in 'BS 5268 - The Structural Use of Timber' are usually adequate, but lower grades tend to be knotty and should be avoided.

A number of practical considerations govern the selection of the size and proportions of solid timber sections for formwork. To prevent lateral buckling under bending loads, timber sections that are relatively tall and narrow require lateral restraint at frequent positions. Figure 2.02 shows this in principle.

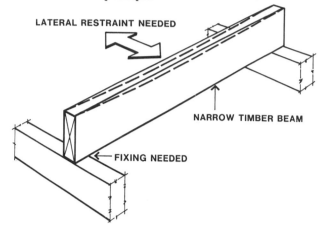

Figure 2.02 - LATERAL BUCKLING OF BEAMS

With wall forms, the nailing of the form face to the studs, and the stud to other framing members achieves this lateral restraint. This is not the case with soffit forms. Frequently, joists are laid loose on the bearers, and the nailing of the plywood to the joists is very limited. As a result there is no effective resistance to the lateral buckling of the top of the joists at mid-span. To limit the tendency buckle laterally, the timber sections should be wide in relation to their height.

Two safety considerations must also be taken into account. Firstly, when soffit forms are being constructed, the formworkers often move about on them. Narrow bearer or joist sections would readily roll over, wide sections would have less tendency to do this.

For the roll-over characteristic alone, it is recommended that the minimum proportions be a depth/width ratio of 2:1 with a minimum width of 45 mm. This information is restated in Chapter 6. The safety aspects of this roll-over problem are well understood in the formwork industry and some formworker use square section joists to minimise the hazards.

Secondly, formworkers may fall when the sections roll, or fall due to other causes. If the narrow joists were lying on their wide face they may not have sufficient strength to carry the impact load of a falling formworker. This consideration will determine a minimum width of joist relative to span.

The size recommendations given here, and in Chapter 6, cannot be taken as a guarantee that the sections will not buckle laterally. The need for lateral restraint is also determined by span, loading and material stress grade. For each case, the formwork designer must determine the lateral restraint requirements.

As noted earlier in the general principles, the cross-sectional accuracy of the timbers is important for the speed of construction, economy and accuracy of the formwork. Unfortunately, sawn timbers are often supplied with unacceptable size variations. To cater for this, it is common practice for timber to be uniformly 'sized' smaller than the nominal size. This can be as much as 5 mm undersize, e.g. nominally 150 mm joists would be sized to 145 mm. This reduction in material should be taken into account in the formwork design calculations.

Accurate sizing is essential in those timbers that directly support the formface material. To have variations in joist or stud sizes would inevitably lead to a wavy line in the finished concrete face. The plywood would deflect under the fluid concrete pressure until it made contact with the face of every member.

Finally, to limit the cost of the stock of formwork materials, the range of timber sizes held in stock should be few as practicable.

2. Timber - Manufactured Sections.

Laminated veneer lumber is a manufactured laminated material of solid rectangular section which is produced by a process similar to that for plywood. Its principal difference to plywood manufacture is that all the veneers are laminated with their grain in the same direction, along the length of the member. The high level of quality control on the process results in a consistent and accurate member of a proven stress grade. The techniques of its use in formwork are similar to that used for sawn solid timbers.

Examples of manufactured sections that are made from plywood webs and solid timber flanges were shown in Figure 2.01. In most cases these are more expensive than solid timbers and are only used in special long span formwork situations. For structural efficiency they have cross-sections that are tall and narrow. Accordingly, they usually require fixing at all supports and the restraint of the compression flange within the span to prevent lateral buckling.

Manufactured sections are widely used on larger wall panel and soffit work because they are very durable and dimensionally stable. Surface coating and the use of laminated plywood or diagonal lattice web plates enable them to resist extremes of humidity with little distortion; furthermore, their bending strength and deflection properties are as good as similar size timbers but at only 70% of the weight. The most common manufactured joist sizes are 200 mm or 240 mm deep, at which sizes dimensionally stable timber sections are very difficult to obtain.

3. Steel.

All the normal hot rolled solid steel sections are used for formwork framing: angles, channels, universal beams and universal columns. Their use is dominantly for long span framing members in large formwork units. All these sections are available in the normal grade and some in higher yield grades.

Hollow sections, round, square and rectangular, are used in formwork. Round tubes are used in the raw form for bracing and, when equipped with suitable end fittings, for soffit supports. Round tube is the base material for the fabrication of most proprietary frame systems, modular support systems and telescopic props. Square and rectangular hollow sections are fabricated into some proprietary products, but their main use is in purpose built formwork.

Folded sheet metal sections have gained wide acceptance in the formwork industry. Standard cold rolled sections, lipped channels, Zeds and top-hat sections, made for general building use, are incorporated in wall and soffit forms as a substitute for solid timber members. Made of high tensile galvanized steel sheet they are usually lighter than the equivalent timber member.

Folded sheet steel members for formwork are not limited to existing industrial products. Special sections can be folded for purpose built formwork.

4. Aluminium.

One of the great advantages of aluminium is its facility to be extruded in an almost limitless range of shapes. The principal restriction is the high cost of the extrusion die. This usually excludes the use of aluminium for special purpose made formwork. However, aluminium proprietary beam sections, and modular soffit systems are marketed. Here the production quantities are sufficient to justify the high die costs.

Aluminium is available in stress grades roughly equivalent to normal structural steel. Unfortunately, aluminium has an elastic modulus approximately 40% that of steel and, section for section, the deflection of aluminium is proportionately greater than the steel beam. Nevertheless, its lightness and durability make it an attractive formwork framing material.

FORMFACE MATERIALS.

The general principles applicable to the selection of satisfactory formface materials were given previously. A number of types of plywood, solid timber and steel sheet can provide suitable formfaces. All of them require the application of release agents to prevent them bonding to the concrete.

1. Plywood.

Plywood suitable for formfaces is available in a wide range of stress grades, veneer arrangements, thicknesses and bond types. It can be used unsurfaced (known as rawform), or purchased with the surface presealed with materials such as phenolics or aluminium. Alternatively, for very high quality surfaces the raw structural plywood can be faced with GRP (fibreglass).

Formwork plywood should have veneer bonding which can withstand at least two years exposure. If a longer re-use life than this is anticipated, such as with forms faced with GRP, plywoods with the fully permanent bond should be used. This is recommended so that the life of the plywood more closely matches the potential durability of the GRP.

Only plywoods that have a known and marked stress grading should be used. The stress grading of plywood is based on the allowable stress values of the particular veneer species and quality. The stress grade defines the basic allowable veneer working stresses and is associated with an elastic modulus. Together these enable a reliable prediction of the strength and stiffness of the plywood sheet.

Thickness-for-thickness there can be a wide range in the available arrangements of veneer numbers and thicknesses. Formworkers need to understand the importance of this. Price alone should not decide purchase decisions. The strength and stiffness of a particular plywood is determined by the thickness and stress grade of its veneers and their arrangement (construction). The construction determines the moment of inertia (I) and the section modulus (Z).

Figure 2.03 - TYPICAL SECTION OF PLYWOOD

So that the outer veneers will always have their grain parallel to each other, plywood is always constructed with an odd number of veneers. A typical section of a plywood sheet is shown in Figure 2.03. Veneers 'A' would all be parallel to one another and veneers 'B' are also parallel to each other but at right angles to veneers 'A'. For most formwork plywoods, these outer veneers have their grain parallel to the length of the sheet.

In determining the bending strength of a plywood only the veneers parallel to the direction of bending are considered. So, for bending along the length of a typical sheet of formply, only the veneers 'A' would be counted. For bending strength across the sheet only the veneers 'B' would matter. Formwork plywood with thick face veneers is very much stronger in the longitudinal direction than the cross direction.

However, if the plywood has thin outer veneers, then the first cross veneers are close to the surface, and the cross strength can be very similar to the longitudinal strength. This is a very useful construction for a formwork plywood. It enables the formworker to cut and fit plywood sheets, without reference to the face veneer grain direction, knowing that it will not be detrimental to the formface strength.

Plywoods with thin face veneers also have advantages when quality concrete surfaces are specified. The thin veneer only responds with a small amount of swelling if there is any moisture penetration. The resultant 'grain imprint' on the concrete face is thus less obvious.

The amount of moisture penetration will be minimised by the surface sealing. For rawform, this will need to be three or more applications of surface sealer and release agent to achieve some degree of success in limiting moisture penetration. Rawform should not be used for forming quality concrete surfaces.

Plywoods with presealed surfaces give the best results, and even these need the site applications of release agent to limit moisture penetration and inhibit adhesion.

For high quality concrete surfaces three combinations of facings and surface veneer have been found to be suitable:

1. A solid face veneer of maximum nominal thickness of 1.3 mm with a surface with a phenolic-type impregnated paper of minimum paper weight of 40 gsm and total weight 120 gsm (40/120).
2. A solid face veneer of maximum nominal thickness of 1.3 mm surfaced with a liquid finish to give a durable wear resistant non-absorbent finish.
3. A solid face veneer of maximum nominal thickness of 1.6 mm (unsanded) or 2.5 mm (sanded) with a surface

with a phenolic-type impregnated paper of minimum paper weight 60 gsm and total weight 150 gsm (60/150).

For general use, where good visual quality is required on work which is to be viewed as a whole, two combinations of facing and surface veneer have been found to be suitable.

1. A solid face veneer of maximum nominal thickness of 2.5 mm with a surface with a phenolic-type impregnated paper of minimum paper weight 30 gsm and total weight 90 gsm (30/90).
2. A solid face veneer of maximum nominal thickness of 3.2 mm with a surface with a phenolic-type paper of minimum paper weight 40 gsm and a total weight 120 gsm (40/120).

For concrete surfaces which are to have an applied finish, such as render or tiles, these two types of facings are also the minimum that should be used. To have less effective surface sealing, will result in hydration staining. As a product of poor hydration, it results in weak concrete and ineffective adhesion of the applied finish. However, applied finishes readily cover minor defects and older plywood, that has had a number of uses, is usually suitable. The less expensive unsurfaced plywood (rawform) can also be used provided that it meets these conditions:

A solid face veneer of maximum nominal thickness of 2.5 mm with the surface well sealed with multiple pretreatment applications and a further treatment before each use with a suitable release agent.

For much structural formwork the finished concrete will be obscured by external cladding, internal linings or earth backfill. Holding tanks, silos, marine structures, bridge soffits and the like, although exposed, are seldom subject to critical examination. In these circumstances unsurfaced (rawform) plywood is cost effective up to about 20 re-uses. Sanded plywood 'good one-side' offers good striking qualities after mould oil treatment, and 'good two-sides', although slightly more expensive, can be turned to prolong its working life.

Lower quality unsanded plywood such as 'sheathing' or 'select tight-face' is used in buried structures where few re-uses are possible.

In wall formwork 19 mm thick plywood is most popular, but thicknesses down to 12 mm are used in soffits and 6 mm in steel frame panels and curved formwork.

The initial cost of 19 mm thick plywood can vary as much as four-fold, depending on quality, finish and coating. Appropriate selection is critical to the overall formwork cost.

Abrasion resistance was listed as one of the desireable characteristics of formface materials. For normal soffit and wall work, the pine faced plywoods will give satisfactory use. They have adequate resistance to normal foot traffic. Surface damage usually only occurs from bundles of reinforcement and pallets of equipment. This can be avoided with appropriate care. However, where there is the possibility of the formface sliding on the concrete surface during stripping and relocation, then harder face materials are needed. Example of this are found with some proprietary mechanised 'climb form' systems.

In the first instance, a harder timber for the outer veneers of the plywood can be used. These are available in the formwork industry. Alternatively, harder plywood facings can be used such as aluminium or GRP. Plywoods faced with aluminium are commercially available. For GRP facings the material must be applied to unsurfaced plywood (rawform).

GRP facing to plywood can be quite costly but it can offer many advantages. Carefully handled, it can give up to one hundred uses before any maintenance is needed. Maintenance involves sanding back and recoating with resin. For more extensive damage, such as that from impact, the affected section of the GRP is cut out and relaid. Careful sanding is needed to hide any differences in texture between the repair and the older GRP facing. The durability of GRP faced forms is more related to the robustness of the formwork structure, than the formface.

It is common practice for the GRP be applied after the raw plywood has been fixed to the formwork structure. The fixings of the plywood to the framing are concealed by the GRP. For success in application of the GRP there must be correct selection of materials and skill exercised in their use. The recommended techniques will be found to vary between resin suppliers but some general parameters can be set.

1. The plywood face veneers must be clean and dry and should have a low resinous content so that there is an effective keying in of the GRP.
2. The formulation of the resin should be selected to suit the intended service life of the formface. This should be discussed with both the plywood and the resin suppliers.
3. The type of glass reinforcement selected must also relate to the intended service life. Normally chopped strand mat with a surface tissue is adequate. If greater strength is needed then woven rovings can be used between two layers of chopped strand mat.
4. The stiffness of the plywood is important. If the vibrator compaction of the concrete induces the formface to vibrate in sympathy, then the resultant concrete face may have serious discolouration. Further, this vibration may cause the GRP to delaminate. The effective plywood stiffness can be increased by framing the form to give a smaller than usual plywood span. Alternatively, a thicker, higher stress grade plywood can be used.
5. The quality of the surface finish of the GRP will depend on the process adopted and the care taken in its finish.

2. Solid Timber.

Although plywood is used to provide the great majority of formfaces, solid timber still has a place. Indeed, in some European countries it is still the only material used. Generally however, its use is now confined to architectural concrete surfaces.

Solid timber can be varied in thickness and sand blasted to accentuate the grain for architectural concrete finishes. Figure 2.04 shows the cross section of an example of this type of formface.

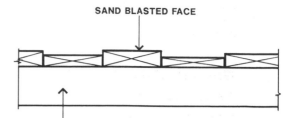

Figure 2.04 - SOLID TIMBER FORMFACE

Wall from Timber Formface Varying in Thickness

As these formface members are spanning between the framing to resist the concrete pressures, as well as forming a decorative finish, they may be quite thick; commonly up to 30 mm thick, and rarely less than 20 mm. The same surface results can be achieved with thinner timbers if they are backed up with a structural plywood.

To provide a satisfactory concrete surface , the timber used should have a tight surface grain to limit moisture penetration and the resultant occurrence of hydration staining. Beyond this, a number of applications of sealants and release agents will usually be required.

A technique, sometimes called 'pre-ageing' can also be used to good effect. This simply involves coating the untreated timber with cement slurry prior to its first use. After drying, the surplus cement is brushed off and the first coat of release agent can be applied. The cement particles tend to fill any open-grained parts of the surface, and are an effective vehicle for the retention of the release agent.

3. Steel Formfaces.

Sheet steel is widely used for the formface of some proprietary modular form systems and special purpose-made formwork. The smooth accurate surface of sheet steel can produce a comparable concrete surface. In carefully fabricated units it is common for the only imperfections to be minor undulations at the positions of the welds.

Steel has the advantages of high strength and a surface hardness that resists rough treatment. Steel is one of the few formface materials that can totally resist direct contact with immersion vibrators. Normal concrete placement rarely damages the surface, but heavy impact, such as dropping the form on a rough surface may dent the face. These dents are often hard to repair as the metal is stretched when dented.

Its principal disadvantage is its weight; modular all-steel formwork panels are made smaller than plywood faced ones to keep their weight within acceptable limits for manual movement. Repairs, if needed, usually involve welding and these facilities may not be on site. It is economical to fabricate steel items, like modular systems, for mass produced items, but expensive for fine detail such as grooves and rebates.

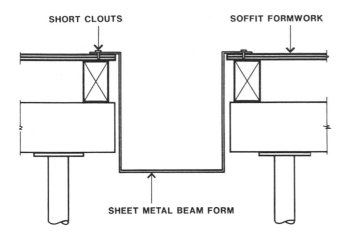

Figure 2.05 - FOLDED SHEET METAL BEAM FORM

Flexibility combined with strength enables steel sheet to be used for formwork in some unique ways. A typical example is shown in Figure 2.05. This gives the cross section of a folded sheet steel beam form which is suspended between two soffit forms. It is fixed in position with short clouts into the soffit formface. Figure 2.06 shows this beam form being stripped after the soffit forms have been removed.

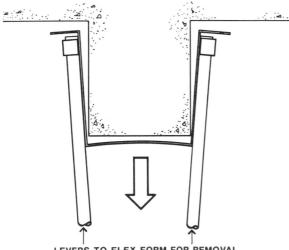

Figure 2.06 - FLEXIBLE STEEL FORM BEING STRIPPED

Levers placed into lugs on the sides of the beam form allow the folded steel sheet to be flexed and removed. The flexibility of modern high tensile sheet steel can be used to advantage in forming many concrete shapes. Columns, pilaster piers on walls, and large groove forms are but a few examples.

4. Fibreglass (GRP)

Fibreglass, glass reinforced plastic (GRP) has already been mentioned as facing material for plywood. It also is a structurally adequate formwork material in its own right. For example, GRP is used to produce purpose made forms for multiple Tee beams and waffle slabs. These are covered in Chapter 7.

However, it is a relatively expensive formwork material, and its use is only justified if there are a number of re-uses, usually more than ten. One of the reasons for this high cost is the slow rate of production. Each of the formwork units has to be 'laid up' in a mould and then cured. Even with accelerated curing techniques this is slow. More moulds can be used to speed production, at additional cost.

One method of speeding production and reducing costs is the combination of GRP with thermoplastics. The outer shell and formface of the product is vacuum moulded in thermoplastic sheet, usually PVC. This is normally a quick and inexpensive process. The large number of plastic moulds can then be internally reinforced with GRP. Provided care is taken to ensure that the heat from the curing resin does not soften the plastic, a very good form product can be produced.

5. Glass Reinforced Cement. (GRC)

This is similar to GRP, except that the main material is a cement sand mix, and the glass fibres are a special alkali-resistant type. GRC is used mainly for permanent formwork. This is covered in more detail in Chapter 9.

6. Rigid Foam Plastic

Rigid polystyrene and polyurethane foams are available in a range of densities. The medium to heavy grades are useful as void forms and penetration forms (holes in floors and walls). Usually they have to be broken out to be stripped and, as a result, are only partly recoverable.

7. Particle Board.

Particle board (chipboard) is made from bonded wood chips. It is available in large sheets with thicknesses from 10 mm to 25 mm. With unsealed surfaces it is quite unsuitable for formwork, the absorption of water is too great. However, specially formulated particle board bonded with moisture resistant glues and surfaced with suitable sealants can be used for formwork. Where these sheets are cut, attention must be paid to edge sealing to prevent water penetration.

RELEASE AGENTS

As noted previously, all formfaces should be given at least one coat of release agent. Its primary function is to prevent the adhesion of the concrete to the formface. Its further functions are the sealing of the formface to limit moisture absorption, and as an aid in preserving the formface and extending its working life.

If adhesion between the formface and the concrete occurs, then one or both of them will be damaged. If, at the time of stripping, the concrete is weaker than the formface then **scaling** occurs with parts of the concrete surface sticking to the form. **Scabbing**, with some of the formface adhering to the concrete, results when the concrete strength is greater than that of the form face.

Common Types of Release Agents.

Neat oils.

These are usually mineral oils; they tend to increase the number of blowholes and their use is usually forbidden.

Neat oils with surfactant.

Neat oils with the addition of a small amount of surface activating or wetting agent assist in minimising blowholes and have good resistance to climatic conditions.

Mould cream emulsions.

These emulsions of oil in water tend to be removed by rain but are a good general-purpose release agent that minimises blowholes.

Water-soluble emulsions.

These produce a dark porous skin that is not durable. They are not recommended.

Chemical release agents.

Consisting of a chemical suspended in a low viscosity oil distillate, they react with the cement to produce a form of soap at the interface. They are recommended for all high quality work. Excessive amounts will cause retardation of the cement hydration. They only require a light spray application. They are more costly than other agents but this is compensated for by the better coverage.

Wax emulsions.

This is a stable wax suspension. They dry off completely and this aids their resistance to climatic conditions.

All of these agents have some effect on the concrete, particularly coloured concrete work. Where coloured concrete work is specified, test panels should be made to check this effect.

Care must be taken in their application. Contamination of reinforcement and concrete surfaces must not occur. Any such contamination can be most efficiently removed by sandblasting. The application of release agent must be adequate to avoid adhesion. However, too much release agent is as bad as too little. With too much it runs to low points and puddles, and can then cause retardation of the cement hydration.

Excessive time of exposure can result in the release agent drying out, running down the formface or being washed away by rain. The shorter the time between application of the release agent and the concrete pour, the better.

FORMWORK FIXINGS

Good formwork construction techniques are aimed at the production of a strong rigid form that has ease of handling, erection, concrete placement and stripping. For a form to repeatedly withstand these vigorous activities, the various components must be adequately fixed together. The fixings must:

1. hold the joints tightly together;
2. fix the formface and the framing members so that the whole assembly becomes a cohesive structure for handling and stripping;
3. enable the easy dismantling of the form and the maximum recovery of material.

In formwork, the fixings such nails and screws that perform these functions are not normally part of the load path that carries the forces from the concrete and concrete placement activity.

In soffit formwork for slabs, the plywood sits on the joists that in turn sit on the bearers. The end reactions of the bearers are resisted by the support system which transmits the loads down to the foundation. Any nailing of the plywood to the framing is usually confined to the plywood sheets at the perimeter of the form. This is done to hold the whole formed area together. This nailing plays no part in the load path from concrete to foundation.

With wall forms, the concrete pressure is carried to the studs which in turn span onto the walers where the loads are carried by the wall ties to the other face of the double faced formwork. The fixings of the plywood to the studs and the studs to the walers are there to hold the form in shape while it is being made, moved, erected and climbed on. These fixings play no part in the load path from concrete pressure to wall ties. This is not to say there is no forces on them. They must be adequate to resist the effects of movement, erection, alignment and stripping.

If movement involves crane hoisting then nailing may not be adequate, extensive screw fixings may be called for. The greater the number of intended re-uses the more attention that will have to be paid to the fixings.

If the fixings are to have a structural function, for example, achieving composite action in a tableform or strengthening a wall form for crane hoisting, then the fixings must be arranged to cater for this situation. Information on the load capacity of the fixings must be sought from the manufacturer so that the formwork designer is able to calculate the type and spacing of fixings.

The fixing requirements will vary from case to case. This book can only give an overview of the more common types and their use. The frequency and type of fixings will depend on the method of use of the form. However, one minimum safety requirement applies to all forms; the fixings must be adequate for the formworkers to be able climb and move on all parts of the form in safety.

Types of Fixings.

1. Nails.

For formface fixing to the framing, and fixings between framing members, nailing is the simplest and most cost effective method. In formwork, nails should only be used to locate the component parts and should not be regarded as having predictable structural characteristics. Three types of nails are in general use: flat head, bullet head and double headed.

Flat headed nails are the most common type used. The large flat head enables a tight secure fixing of the formface to the framing. However, the reflection of the nailhead is often visible on the concrete surface. For quality concrete surfaces this is usually unacceptable. In these cases bullet head nails can be used. These are not as effective as the flathead nails in clamping the formface to the framing, and under load, the smaller head of the bullethead nail may be pulled down into a plywood formface.

For plywood formface fixing, a practical recommendation is to use nails of 2.5 mm minimum diameter and longer than 2.5 times the plywood thickness, 50 mm for 17mm plywood. Nail sizes and lengths used for framing fixings will depend upon the particular case.

Double headed nails are specifically made for the formwork industry. They have a second flat head approximately 10mm down the nail shank from the end. As shown in Figure 2.07, when driven, the second flathead provides the tightening action on the surface of the timber. The top head of the nail protrudes for easy nail withdrawal.

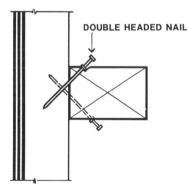

Figure 2.07 - DOUBLE HEADED NAILS IN USE

Double headed nails are only used between framing members and for fixing wedges. They are quite unsuitable for formface fixings. In addition, they are significantly more costly than conventional nails.

2. Screws

Where fixings more robust than nails are needed, screws can be used. These can be wood screws, power driven screws or coach screws.

Hand driven wood screws, countersunk or round head have largely been superseded by power driven screws for formwork. Power driven countersunk or buglehead screws are used for plywood fixing to the timber or steel framework. They can also be used for fixings between framing members, but this is often better done with hex-head screws. Hex-head screws are also used for fixing steel framing and fittings to timber.

Coach screws, both hexagonal and square head, are used when the strongest screw fixings to timber are needed. They are normally used for fixing steel framing or fittings to timber.

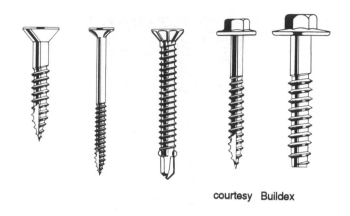

courtesy Buildex

Types of Screws Used in Formwork

3. Bolts.

Both mild steel and high strength bolts are used in formwork framing. These can be designed as part of the load paths discussed previously. Bolts should be hot dip galvanised for longevity and ease of removal. When used with timber framing, large square washers should be fitted to minimise the crushing of the surfaces of the timber.

High strength bolts are confined to metal-to-metal joints. Their most common use is found in achieving a reliable friction grip at adjustment points. An example of this is given for column forms in Chapter 5.

PROPRIETARY DEVICES AND COMPONENTS.

As will be shown throughout the following chapters the formwork industry is well served by suppliers of proprietary formwork components and systems. These are in addition to the normal industrial products that are used in conventional formwork.

Only a few of these proprietary formwork products are expendable like the two column forms shown above. The remainder are re-useable and designed for a long working life. They range from individual components, such as

Proprietary Quick Strip Soffit Form System

Proprietary Plastic Column Form

Spirally Wound Proprietary Column Form

acceptable wear, when to reject), minimum bearing areas, anchorage requirements, bracing requirements,

Other technical information, although not always needed, should be available. This includes:

1. Whether the data was test derived or calculated,
2. If test derived, data on the method and arrangement of the test, place and dates of the tests, name of certifying authority,
3. If calculated, information on the controlling assumptions used in the design calculations.

props and wall ties, that are incorporated into conventional forms, through to complete modular wall, column and soffit formwork systems. Examples of these are given in the relevant chapters.

So that formwork designers can evaluate the suitability of proprietary components for particular formwork projects they need to be adequately informed. At a minimum this information, where relevant, should include:

1. An illustrated description of the item,
2. A list of the items of equipment of the type which are available, range of sizes, part numbers, overall dimensions, and information on cambers,
3. A description of its functions,
4. Its self weight,
5. Data on is relevant working load capacities and information on its limit state capacities,
6. Instructions on its use and the safe limits of its use, points requiring special attention, parameters which indicate that maintenance is required (limits of

CHAPTER 3: GROUND FORMS

Although concrete work in the ground is usually hidden from view, the need for accuracy in the formwork associated with it should not be treated lightly. The lines, levels and positions of the features of footings set the starting accuracy for following work such as walls and columns. Careless work at this first stage, creates the later problem of correcting levels and alignments to reach the standard required for the exposed concrete work. This needless expense should be avoided.

Work in the ground can be difficult. The workspace is often very confined, and it can be almost impossible to keep the excavations, reinforcement and the formwork clean, particularly in wet weather. Where the ground is hard, it can be difficult to drive pegs for the support of the forms. When the ground is very soft, even quite long pegs, driven deeply, may not give the needed support.

It is common practice for footings to be poured against the faces of the excavation, so the plan size of the excavation is usually the net size of the footing; no extra excavation being done to provide access and workspace for the formworker. As a result, forms for the tops of the footing and for the support of reinforcement are often suspended down from the ground surface.

The suspended forms are usually carried on beams bridging to pegs driven into the ground. If these pegs are too close to the face of the excavation, the act of driving the pegs and the load on the pegs may make its banks unstable. To avoid cave-ins, the pegs should not be close to the excavation even in firm ground. With soft ground, extensive shoring may be needed for the security of the banks and the safety of the workplace.

The design and construction of shoring systems for the security of excavated areas requires an understanding of soil and rock mechanics. Further, the behaviour of soils can be seriously affected by rain and flooding. This aspect of construction work is outside the scope of this book.

Formwork on the ground surface can also have construction problems. The ground may be rocky and reasonable peg penetration and alignment hard to achieve. Like formwork in excavations for footings, wet weather can make work difficult and dirty.

This topic of forms at or in the ground will be dealt with in two broad areas, forms on the surface such as edge forms and, formwork for footings in excavations. Other work in excavations, for example formwork for basement walls, is covered in later chapters.

EDGE FORMS

The basic requirements of an edge form is that it should be straight, strong and stiff. Straightness is needed to achieve an accurate line. Adequate strength is required to resist the loads that act on it. Concretors walk on them, screeds, often heavy vibrating screeds, use them as a guide and the concrete pressure pushes outwards. Under these loads an accurate line and plumb face must be maintained. For this they must be stiff.

Where concrete depths are shallow, say up to 200 mm, solid timber of 50 mm width is suitable.(Figure 3.01(a)) For greater depths, forms can be fabricated with plywood faces and stiffening plates of at least 75 x 50 at the top and bottom and widely spaced studs of the same size.

The edge forms span continuously along a line of pegs. The pegs maintain the line, level and plumb of the forms, and transmit the loads to the ground. Pegs are usually cut from 50 x 50 timber, their spacing depending on the edge form construction, the depth of concrete and the soil characteristics. For soft soils the peg spacing may need to be as close as 600 mm. In harder clays the spacing may be quite satisfactory at 1.2 metres.

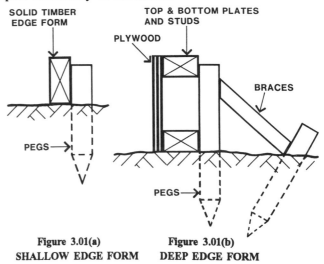

Figure 3.01(a)
SHALLOW EDGE FORM

Figure 3.01(b)
DEEP EDGE FORM

Shallow edge forms, say for 100 mm slabs, rarely require bracing. The deeper edge forms, say 300 mm and over, on firm soils, or 200 mm on soft soils, usually require bracing to the pegs. Figure 3.01(b) includes an example. In workable soils, the pegs can firstly be driven to approximate line and depth. A small movement sideways for accurate peg alignment can usually be achieved by compacting the surface of the ground on the appropriate side of the peg. This is usually done with a sledge hammer.

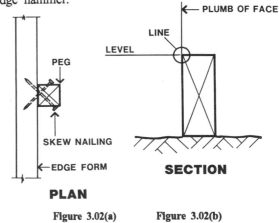

PLAN

SECTION

Figure 3.02(a)   **Figure 3.02(b)**

The edge form can then be fixed to the pegs by skew nailing. (Figure 3.02(a)) The aim is a plumb faced edge form set true to line and level. (Figure 3.02(b))

Keyed Edge Forms

Heavy moving loads, such as fork-lift trucks, often cause excessive vertical movement between adjacent industrial floor slabs. To prevent this relative movement, keyed joints are often specified. (Figure 3.03)

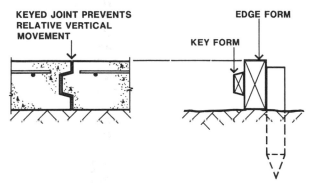

Figure 3.03 - KEYED JOINT AND EDGE FORM

Stripping a keyed edge form is usually a two stage process, first the edge form and then the key form. To aid this, the key form is only lightly fixed to the edge form. When the edge form is stripped, the key form usually remains in place. Although the shape of the key is tapered, its form does not readily strip out. The key form usually requires a further means to ease its stripping.

For a timber key form, a saw kerf is cut in the back. With sheet metal or plastic forms they are made of material thin enough to permit flexibility. (Figure 3.04) In both cases the key form can be slightly squashed for ease of removal.

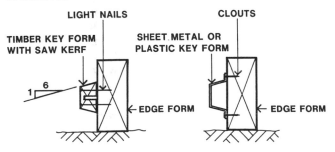

Fig 3.04 - KEY FORMS FOR EASE OF REMOVAL

Dowelled Joints in Slabs

Another method of inhibiting relative vertical movement between adjacent slabs, is the dowelled joint. It also permits horizontal thermal expansion and contraction in the slab joint.

It is important that the steel dowel bars be kept in line and parallel to each other. This is needed in the first instance for ease of stripping the edge form, and later for the efficient operation of the expansion joint.

An effective method is to install a timber alignment strip pegged to the ground at a line and level to suit the dowels. The steel dowels are held by nailing to the strip and passing through holes drilled in the edge form. (Figure 3.05(a))

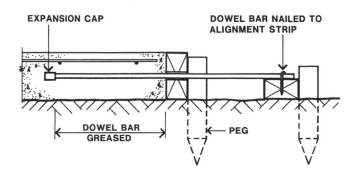

Figure 3.05(a) - DOWELS NAILED TO TIMBER STRIP

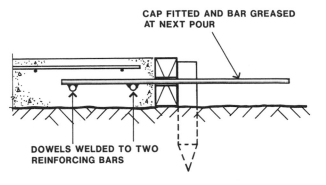

Figure 3.05(b) - DOWELS WELDED TO BARS

Figure 3.05(b) shows another effective alignment method. The dowelled bars are held parallel, and in line, by being welded in sets to two long reinforcing bars. Attention still needs to be paid to aligning the dowels bar sets, but the task is easier than with loose dowel bars.

Continuity of Reinforcement

For construction joints in slabs and crack-control joints, continuity of the reinforcing mesh is often required. In these cases a two piece edge form can give effective results.

For mesh reinforcement a simple two part edge form can be used. (Figure 3.06) As the longitudinal wires of the mesh are usually small, two plain rectangular pieces of timber can be used and the grout loss past the wires will usually not be significant. The lower part of the edge form is fixed to the pegs, set with a small gap underneath it to ease its stripping. The upper part of the edge form is nailed to the lower.

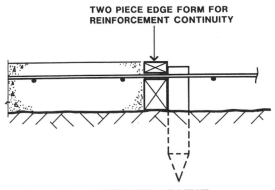

Figure 3.06 - CONTINUITY OF MESH

For reinforcing bars, 12 mm and larger, the edge form can be drilled and then saw cut on the centreline of these holes to split the form in two. The holes should be drilled 4 mm oversize for ease of bar installation. (Figure 3.07)

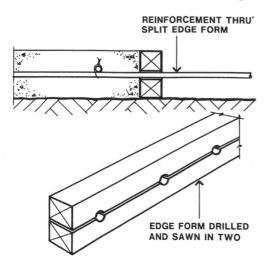

Figure 3.07 - SPLIT EDGE FORM

The lower part of the edge form is nailed to the pegs. The reinforcing bars are then placed and the top part of the edge form nailed on. On stripping, the small amount of slurry that penetrates around the bars and between the two parts of the edge form is readily cleaned off. As before, a gap below the lower part aids stripping the lower part of the edge form. This type of form is also suitable form dowelled joints.

Waterstops in Edge Forms

Waterstops are frequently used in joints in basements, swimming pools, water tanks and stormwater channels. A two part edge form is an effective way to hold the waterstop in place. (Figure 3.08)

To fit the waterstop, both parts of the edge form should be rebated to suit the profile of the waterstop. This effectively limits horizontal movement of the waterstop during concrete placing.

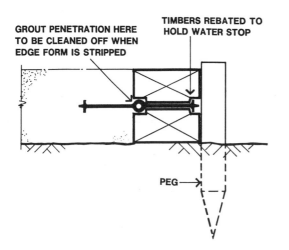

Figure 3.08 - EDGE FORM WITH WATERSTOP

The outstanding part of the waterstop, which is to be embedded in the concrete, should be supported on its outer edge to stop it curling over during the pour. Wiring to the top reinforcement is one common method.

During concrete placement, a small amount of grout penetrates around the waterstop into the rebates in the edgeform. However, this is easily broken away after stripping.

Pegging Edge Forms on Hard Ground

Often hard or rocky surfaces cannot be penetrated by sharpened timber pegs, pine or hardwood. Further, when steel pegs are used it is usually difficult to control the accuracy of their position due to rocks in the ground. Figure 3.09 shows a method that caters for this.

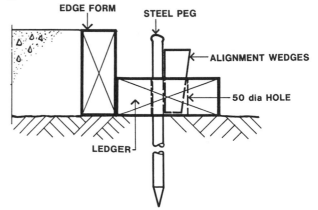

Figure 3.09 - EDGE FORMS ON HARD GROUND - 1

The edge form is nailed to a continuous ledger which is bedded continuously on the ground. The steel pegs (usually 20 mm dia.) are driven into the ground through the large holes (50 mm dia.) in the ledger. The aim is to place them centrally in the holes. However, the rough ground often forces the pegs off position and the oversize holes cater for this.

After pegging the edge form can be aligned and its position fixed by wedging between the steel pegs and the appropriate edges of the holes.

It is best to have the 50 dia. holes in the ledger at quite close centres, say 300 mm, even though only every third one will be used. If penetration of the ground is impossible at one hole then the next one can be tried.

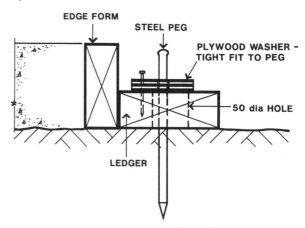

Figure 3.10 - EDGE FORMS ON HARD GROUND - 2

An alternative to wedging at the holes is shown in Figure 3.10. Tightly fitting plywood washers are placed on the steel pegs before they are driven. After driving the pegs, the edge form is aligned and the plywood washers then nailed to the ledger to fix its position.

Moisture Barrier Laps at Edge Forms

Many slabs are specified to be placed on waterproof membranes, also called moisture barriers or vapour barriers. To enable an efficient lap joint to be made in the membranes under adjacent slab pours, it is usually necessary to extend the membrane beyond the slab joint. To maintain watertightness, it is common for the specification to prohibit the piercing of the membrane by pegs for the edge form.

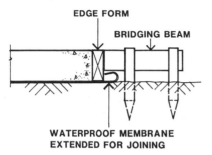

Figure 3.11 - CANTILEVERED EDGE FORM

Figure 3.11 shows one effective method of achieving this. The edge form is seated on the waterproof membrane and held to line by cantilever bridging beams, which are fixed to pairs of pegs driven into the ground beyond the membrane. The spacing of the bridging beams is similar to the peg spacing discussed previously. As will be shown below, the support of edge forms on this cantilever arrangement is a common technique.

Cantilever Supports to Edge Forms

The cantilever arrangement also provides a solution to forming a step in the top of the slab near its edge. Figure 3.12(a) shows such a case. The pegs at the edge form are extended upwards to provide support for the cantilevering bridging beams which carry the step form. They are tied down at their outer ends by deep pegs. Note that when concretors stand on the step form, these outer tie-down pegs will be subject to considerable uplift.

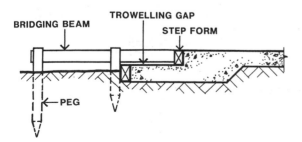

Figure 3.12(a) - TOP STEP IN SLAB

As Figure 3.12(b) shows, a double bridging beam arrangement is needed to support the step form around a corner. Both figures show the underside of the bridging beam located above the slab surface. This gap (usually 25 mm or more) gives access for trowelling the surface. The

trowel access at the step form can also be improved and details are given in Chapter 8 STAIR FORMS.

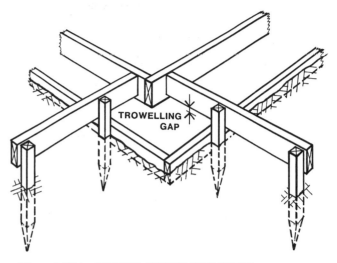

Figure 3.12(b) - CORNER OF TOP STEP FORM

The examples of edge forms shown above, have all related to the simple cases of edge forms for slabs placed on a generally level surface. As discussed at the start of this chapter, where the edge forms are required at the face of an excavation, the problem of cave-in from peg driving usually occurs. A simple example is the construction of a baffle wall under the outer edge of a slab.

These baffle walls are sometimes specified to inhibit the movement of soil moisture or prevent the intrusion of vermin under the slab. If the baffle wall is specified to be poured integrally with the slab then a cantilevered edge form like that shown in Figure 3.13 is needed. This has the peg far enough away from the face of the excavation for the baffle wall, that it will not cause a cave-in.

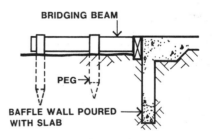

Figure 3.13 - BAFFLE WALL POURED WITH SLAB

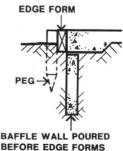

Figure 3.14 - BAFFLE WALL POURED FIRST

However, a change in the construction procedure can give significant savings. If the baffle wall can be excavated and poured before the slab, the edge form construction need not commence until the concrete of the baffle wall has set. Then a simple edge form is all that is required. (Figure 3.14)

For deeper excavations, and those that require a large part of the edge to be formed, fabricated edge forms of plywood and timber are usually needed. Bracing of the edge form may also be needed to prevent twisting of the face as the concrete placement progresses. The case shown in Figure 3.15 includes a rebate form. Such rebates are often specified for brick construction.

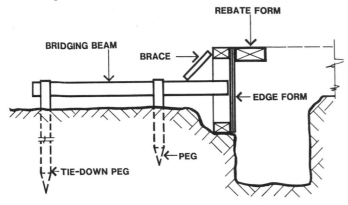

Figure 3.15 - DEEP EDGE BEAM WITH REBATE

Special provision has to be made to support the corner of cantilevered edge forms. Figure 3.16 shows a method that extends one edge form beyond the corner to a support peg. The other edge form is cleated onto this extended edge form.

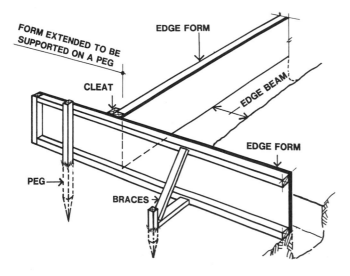

Figure 3.16 - DEEP EDGE FORMS AT THE CORNER - 1

Another method is to carry the corner of the edge form on a diagonally placed bridging beam which is carried on support pegs which are placed well away from the excavation. (Figure 3.17) Hangers from this beam should extend down to the bottom plates of the edge forms.

This second corner detail involves suspension of the formwork. This is very common in formwork in the ground. More examples will be shown later in this chapter.

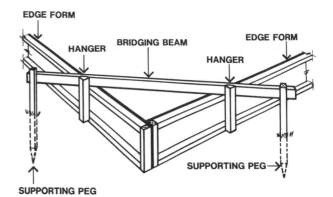

Figure 3.17 - DEEP EDGE BEAMS AT THE CORNER - 2

FORMS FOR FOOTINGS

Isolated Rectangular Footings

For simple isolated rectangular footings, with pedestals (plinths) to carry steel columns, it is common practice to pour the base first, and after that concrete has set, to form and pour the pedestal. Figure 3.18 shows this second stage.

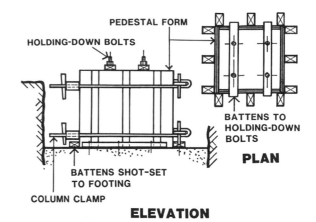

Figure 3.18 - PEDESTAL FORMWORK

The pedestal is formed as a short column form complete with column clamps. (Refer to Chapter 5 COLUMN FORMS) This pedestal form can be located by battens shot-set to the concrete base on all four sides. If this shot-setting is not permitted, the pedestal form can be held in position by strutting it off the excavation.

Holding-down bolts, for the steel column which is to be erected later, are suspended within the pedestal form from battens fixed to the top of the form. It is common for the holding down bolts to be welded into a cage to ensure that they remain plumb during the concrete placing.

When it is specified that concrete to the base and the pedestal are to be placed integrally, then the pedestal form must be suspended in position. (Figure 3.19) The pedestal form is shown suspended from bridging beams which are carried on pegs. To avoid caving-in the face of the excavation, the pegs must be located well back from the faces of the excavation. Braces should be fitted in two directions to prevent sideway of the form, and Kentledge should be added to prevent flotation.

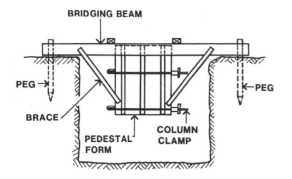

Figure 3.19 - INTEGRAL POUR OF BASE AND PEDESTAL

Reinforcement Support

Even where two stage pours are possible, bridging beams are often used to suspend reinforcement for the first stage of the pour. Figure 3.20 shows a wide strip footing with vertical reinforcement from this base up to the slab. The bridging beams carry a timber member which supports the reinforcement. A similar technique can be used to carry starter bar cages for pedestals and columns.

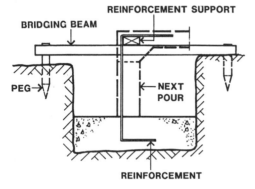

Figure 3.20 - REINFORCEMENT SUPPORT

Strip Footings

Steps in strip footings, particularly those for shallow strip footings, can also be formed by suspended formwork.

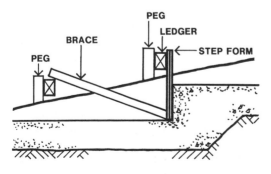

LONGITUDINAL SECTION

Figure 3.21(a) - STRIP FOOTING STEP

Figure 3.21(a) shows a longitudinal section of a step in a shallow strip footing. Figure 3.21(b) shows a perspective view. The stepform, usually just a single piece of plywood, is suspended from a ledger (bridging beam)

carried on pegs. The bottom of the form is braced back to another pegged ledger.

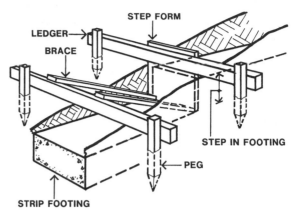

Figure 3.21(b) - FOOTING STEP - PERSPECTIVE VIEW

For strip footings in deep trenches it is often simpler to hold the step formwork in place by wedging it off the sides of the trench. This can be quite effective for small steps.

Foundation Beams to Eccentric Footings

A more complex case is the formwork to the stabilising beam of an eccentrically loaded footing, sometimes called a 'pumphandle footing'. (Figure 3.22)

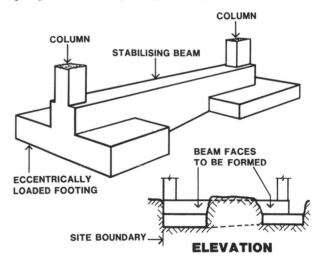

Figure 3.22 - ECCENTRIC FOOTING AND BEAM

Usually the whole unit, footings and beam, must be poured in one operation. While it is quite common to place the concrete of the centre section of the stabilising beam against the excavated faces, the beam faces immediately above the footing pads must be formed.

Figure 3.23 gives a cut-away perspective view of the suspended beam-side formwork above the footing pads. The forms are hung from bridging beams. They must be strongly braced to resist the lateral fluid concrete pressures from the concrete in the beam. The pegs carrying the bridging beam must placed well back from the edge to avoid cave-ins.

The beam side forms need to be joined to the faces of the excavated trench sides to prevent concrete flow and loss around their ends. The plan view shown in Figure 3.24 illustrates one method.

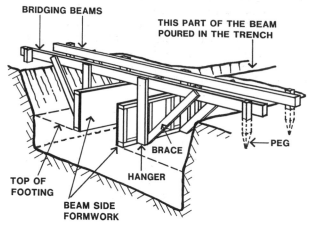

Figure 3.23 - ECCENTRIC FOOTING - PERSPECTIVE VIEW

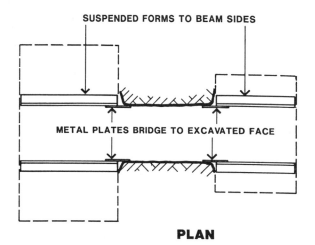

PLAN

Figure 3.24 - JUNCTION OF TRENCH SIDES AND FORMS

A metal plate is bridged across the gap between formwork and excavated face. If the gap is narrow, say 25 mm or less, then sheet metal as thin as 1.6 mm may be adequate. This plate should be regarded as 'lost' or 'sacraficial' formwork as its recovery for further use is usually difficult.

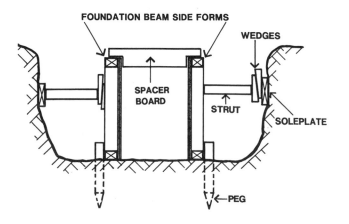

Figure 3.25 - FORMS TO CENTRAL BEAM

If it is decided to over-excavate the beam in the centre section, then the beam sides must be formed. A sectional view is shown in Figure 3.25. Here the base of the beam is poured on the ground, (or a void former if specified - see Chapter 9) and the side forms are pegged at the base. Spacer boards hold the top of the sides apart prior to the concrete placement, and these sides are strutted off soleplates on the excavated trench sides.

Once the concrete placement and compaction is complete, the spacer boards can be removed for ease of screeding and finishing the concrete surface. Figure 3.25 shows a relatively shallow beam. For deep beams another row of lateral strutting will most probably be required along the bottom of the forms.

KICKERS

At the beginning of this chapter it was stressed that concrete, which is the base for following work, must be accurate. One of the methods of making an accurate start to walls and columns is to construct 'nibs' or 'kickers'. Figure 3.26 shows that a kicker serves three functions.

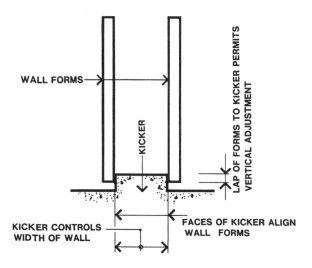

Figure 3.26 - THE FUNCTIONS OF KICKERS

The faces of the kicker give alignment to the wall forms and control the width of the wall. The height of the kicker permits adjustment of the level of the wall formwork. This last function is vital where the wall form has visual details, such as horizontal groove forms that must be set accurately to level.

In addition to the three features noted above, a kicker is a means of achieving a grout tight joint at the interface between concrete and formface. Without the kicker the formwork would, in most cases, be seated on the top of the footing, a surface that is often rough. Grout loss usually results.

However, it should be noted that the erection of wall forms does not always require a kicker. Some examples of other methods of controlling the position of wall forms are described in Chapter 4 WALL FORMS.

A kicker also forms a construction joint and so is often detailed on the project documentation. Kickers detailed with a groove have two main problems. Firstly, as shown on Figure 3.27(a), the access for pouring concrete is very restricted. Vertical reinforcing bars further restrict the concrete placing access.

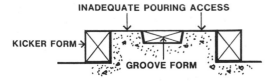

Figure 3.27(a) - KICKER WITH GROOVE

Secondly, the achievment of a grout tight joint and sufficient clamping action to hold the forms in place requires considerable tension in the wall formwork tie-rods. (Refer to Chapter 4 for details) As indicated in Figure 3.27(b) this can lead to cracking of the kicker. Ideally, the kicker should have a solid cross section.

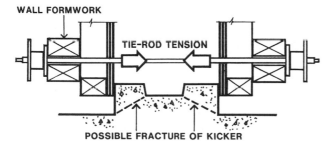

Figure 3.27(b) - KICKER LIABLE TO FRACTURE

Figure 3.28 shows a simple case of forms for a kicker on the surface of a wide footing, such as that for a retaining wall. Pegs supporting the edge forms are extended to carry bridging beams, from which are suspended the kicker forms. The bridging beams also provide support for the reinforcement starter bars. For wide footings, scaffolding planks for worker access can also be carried on the bridging beams.

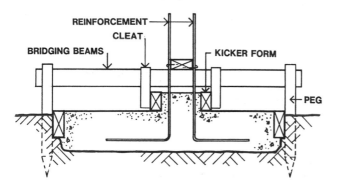

Figure 3.28 - KICKER IN THE MIDDLE OF A FOOTING

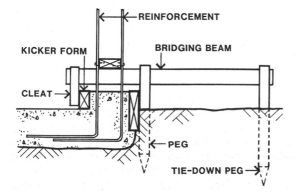

Figure 3.29 - KICKER AT THE EDGE OF A FOOTING

Forms for a kicker at the edge of a footing or slab is shown in Figure 3.29. The principles are the same as the previous case but particular attention must given to getting sufficient tie-down capacity in the outer pegs.

Figure 3.30 is another example of forms for a kicker at the edge of a slab. In this case the inner face slopes up from the top of the slab, and then forms the normal parallel sided kicker. This detail is common in water retaining structures.

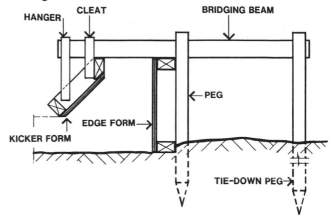

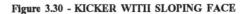

Figure 3.30 - KICKER WITH SLOPING FACE

The suspension techniques described earlier can also be used to form kickers on footings which are in deep trenches. Bracing is essential to prevent side-sway movement of the kicker forms. (Figure 3.31)

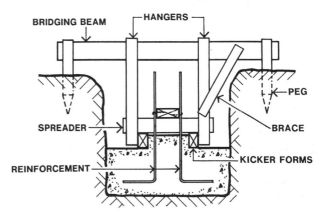

Figure 3.31 - SUSPENDED KICKER FORMS

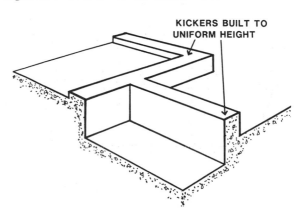

Figure 3.32 - KICKERS BROUGHT TO A UNIFORM HEIGHT

Kickers can also be used to control the start levels of wall forms. Figure 3.32 shows that kickers can be built to a uniform height to give one starting level for walls. This enables wall forms of uniform height to be used.

In the second example, Figure 3.33, the kickers to a sloping footing are used to bring the junction with the wall forms to a series of level lines. Once again, the aim is to simplify the later wall forming operations.

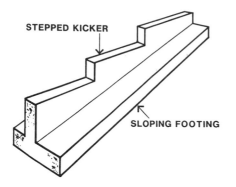

Figure 3.33 - KICKERS ON A SLOPING FOOTING

A simple method of constructing these involves kicker side forms of uniform height constructed on the slope on top of the footing. (Figure 3.34 - ELEVATION) Vertical step forms are installed between the side forms. With adequate fixings they act to set the side forms parallel and maintain their alignment.

These deep kicker side forms require adequate bracing to resist the concrete pressures. Struts to the trench sides, with wedges for tightening are the normal practice. (Figure 3.34 SECTION) In cases of very deep kickers they may need to be tied like wall forms. (Refer to Chapter 4 Walls)

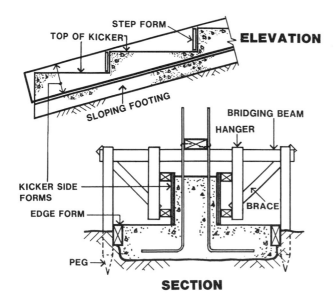

Figure 3.34 - KICKER FORMS TO SLOPING FOOTING

Another type of construction work that needs kickers are tanks in the ground. Figure 3.35 shows the cross section of a type of rectangular tank often constructed in

waste water treatment work. Only the inner face of the tank is to be formed in this case. A construction joint permits the pouring of the base with kickers at the walls. Formwork for a rectangular central drain will have to be provided.

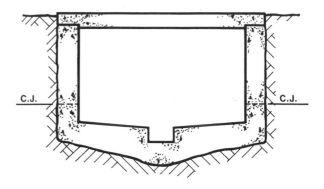

Figure 3.35 - CROSS-SECTION OF IN-GROUND TANK

A method of forming the base is shown in Figure 3.36. Forms for its features, side kicker forms and the central channel form, are suspended from bridging beams which are carried on pegs at the ground surface. The pegs must be well away from the edge of the excavation to avoid cave-ins.

The position of the hanger assembly is stabilised by fixing to horizontal timber struts which are tightly wedged to the sides of the excavation. These struts must be securely fixed to the hangers. The formworkers who build the form and the concretors who follow them will all climb on the bridging beams, struts, kicker forms and channel forms. Some of these members will have scaffold planks placed on them. Secure fixings are vital; bolting would be the most effective.

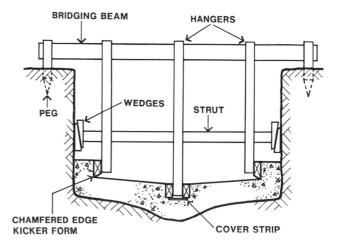

Figure 3.36 - FORMWORK FOR BASE OF TANK

The kicker forms are fixed to the hangers and are shown with a chamfered inner edge for trowel access. (Other methods of achieving trowel access are given in Chapter 8 STAIR FORMS.)

The central channel is formed from two timbers, one to each side of the hanger. The bottom gap between the two timbers is closed over with a cover strip. This can be thin plywood or sheet metal. The channel form is stripped by removal of the hangers, followed by the two longitudinal timbers. The cover strip is usually destroyed during stripping.

****

The information in this chapter has been confined to formwork that is recovered at stripping. Permanent (sacraficial or lost) forms are also used extensively in forming concrete on or in the ground. Void formers under slabs and beams are one example. This topic is covered Chapter 9.

CHAPTER 4: WALL FORMS

Essentially, there are two types of wall formwork. One type is for walls requiring only one face of formwork. Examples of this group are walls for the basements of buildings, rectangular tunnels, tanks in the ground and formwork for mass concrete dams. (Figure 4.01) The methods by which the forces on these forms are catered for are discussed in the last topic in this Chapter.

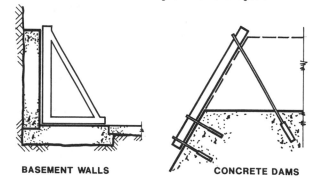

BASEMENT WALLS   **CONCRETE DAMS**

Figure 4.01 - SINGLE FACED WALL FORMS

The other broad class is double faced formwork for free-standing walls. Common examples of this are retaining walls and service cores for buildings. These walls may be poured in a single operation or in a sequence of pours. This latter case will call for construction joints between adjacent pours.

Such sequential work can all be on the one level with each pour being an extension of a previous one. This was briefly discussed in Chapter 1. Alternatively it may be a vertical sequence as encountered in multi-storey work. In both cases the walls may consist of straight sections or may be curved in plan.

WALL FORM LOADING

To devise satisfactory formwork it is necessary to have a detailed appreciation of the loads and combinations of loads that can act on the formwork at the various stages of its use. As discussed in Chapter 1, the construction procedure can be classified as a three stage sequence. During all three stages, safety is a paramount consideration.

STAGE 1 encompasses all the activities that occur prior to the placement of the concrete. This includes the fabrication of the formwork units, their hoisting into place, their temporary bracing during the fixing of reinforcement and completion of erection of the forms, the completion of access platforms and ladders, and the final bracing and adjustment to line, level and plumb.

After fabrication, the wall formwork has to be hoisted from the horizontal position to the vertical. A similar situation may occur when wall formwork is brought into use from its storage position. If crane handled, the hoisting and slewing of the form can induce considerable bending

in the form (Figure 4.02). The framing of the form and all its connections must be adequate to provide sufficient stiffness to cater for this.

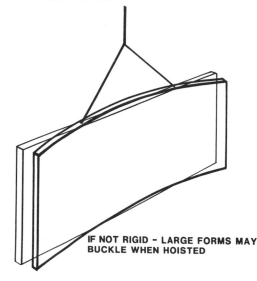

IF NOT RIGID – LARGE FORMS MAY BUCKLE WHEN HOISTED

Figure 4.02 - FORMWORK BENDING DURING HOISTING

When in position, the wall form must be safely braced before it can be released from the crane. If only braces to the top of the wall form are installed and the bottom is not secured then the problem shown in Figure 4.03, may occur with a gust of wind.

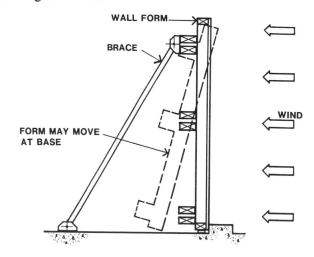

WALL FORM

BRACE

WIND

FORM MAY MOVE AT BASE

Figure 4.03 - INSTABILITY DUE TO WIND

Similar dangerous movement may occur if the wall forms are struck by other components when those are being crane hoisted.

When the wall form unit is used in a climbing sequence, adequate fittings are required to control, and later adjust, its position. Safe fixing is needed for attachment to the previously poured wall. (Figure 4.04).

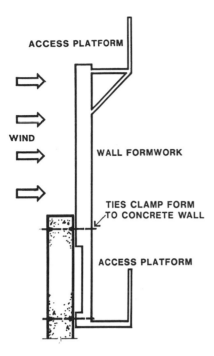

Figure 4.04 - CLIMBING FORM FIXED TO WALL

The concrete of this lower wall must also have developed sufficient strength before the formwork can be safely fixed to it. The top of the concrete wall is subject to the weight of the form as well as the combined bending effect of wind on the formface and the eccentric load action of the formwork and loaded access platforms.

STAGE 2 is the act of placing the concrete. The factors that determine the pressures on wall formfaces were discussed in Chapter 1. For normal pouring procedures, the upper limit of pressure is the hydrostatic head related to the pour height and the concrete density. If setting or partial setting of the concrete occurs before the pour is completed, there will be a reduction in the rate of build up of pressure on the formface.

The factors controlling the concrete setting are the vertical rate of pour, the concrete temperature and the characteristics of the concrete. The plan size of the wall being poured may cause an increase in pressure by reflection of vibration. (refer to Figure 1.16)

Dropping the concrete has an impact effect that can increase the effective pressure on the form. Figure 4.05 illustrates an example where the concrete is dropped from a point higher than the top of the pour. Here a pair of large forms are being used to form a short wall.

Figure 1.23/1 showed one case of possible impact between construction equipment, and the formwork assembly which must be allowed for. Similar impact can occur with concrete boom pumps and hoses. The collision can occur with the braces, the access platforms or the forms themslves. If this happens there will usually be a misalignment of the form, and perhaps even its failure.

It is unreasonable to expect the formwork to be able to resist the full range of impact forces. But, the formwork should be so constructed that impact does not totally collapse the forms and endanger workers and the public by the generation of debris.

Wind and access platform loading can also occur simultaneously with the STAGE 2 activity of concrete placement. This platform loading can be quite high with concrete placers, their vibrators, the pump boom operator and supervisors all on the platform at the same time.

It is essential that the formwork be closely observed during concrete placement. The concretors doing the pour will be concentrating on their work and not on the state of the formwork. Separate personnel, 'formwatchers', will be needed for this. Their job will be to detect the effects of any overload and take immediate action.

The first part of **STAGE 3** is that period when the formwork supports and protects the wall, while the strength of its concrete develops. The formwork loads during this phase are usually limited to wind and construction activity. The second part of this final stage starts with formwork removal. This is delayed until the concrete is strong enough for the wall to safely freestand and its surface is hard enough to resist damage during this stripping.

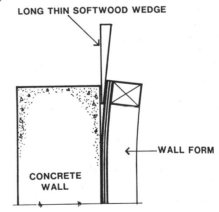

Figure 4.06/1 - STARTING WALL FORM STRIPPING

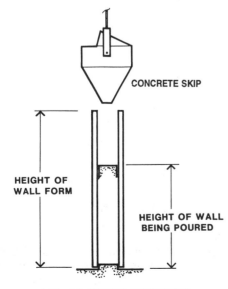

Figure 4.05 - DROPPING CONCRETE

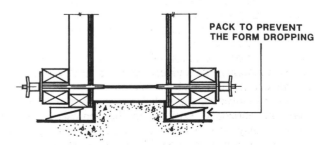

Figure 4.06/2 - SUPPORTING THE BASE OF THE FORMS

Ideally, formwork should be removed at right angles to the concrete face in one operation. This is usually difficult to achieve, and an acceptable compromise is a peeling action. Sometimes the formface does not readily move away from the concrete. As shown in Figure 4.06/1 this can usually be initiated by inserting a long thin softwood wedge at a top corner and then progressively prising the top edge away. The base of the form should be supported to prevent dropping. (Figure 4.06/2) Care must be taken to avoid any sliding. Sliding would deface both the formwork and the concrete surfaces.

The final part of STAGE 3 for wall forms is the moving of the formwork unit. This can be movement to a storage location or to a position for its next use. In this latter case it is, strictly speaking, the start of the next STAGE 1.

DOUBLE FACED FORMS

At its simplest, a double faced formwork assembly will be a box consisting of two opposing faces tied together on some regular pattern, and with the ends closed. In conventional formwork the formface is usually plywood with a sealed face. In some cases the formface may be solid timber boarding. The formface spans onto a grid of timber framing members to bring the loads to the cross wall ties.

In most of the examples that follow, the framing is shown as solid rectangular timber sections. However, in many cases other members such as cold rolled sections, hot rolled sections, rectangular hollow sections or round tubes can be used. Examples are shown in Figure 4.07. The cold rolled sections can be lipped channels or zed sections, either singly or in pairs. Hot rolled sections can be angles, channels or I sections,

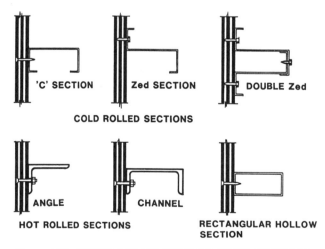

Figure 4.07 - TYPICAL WALL FORM FRAMING MEMBERS

Broadly, there are two types of wall formwork, identified by the direction of their framing: horizontal waler and vertical waler forms. Figure 4.08 gives the cross section of a horizontal waler form. For this type of form, the vertical studs are fixed directly to the back of the formface material, in this case plywood. At the ends of the studs, horizontal top and bottom plates are fixed. Wedges under the bottom plate hold the wall forms to level.

Pairs of horizontal walers are fixed to the outer face of the studs. The wall ties connect between the walers on opposing wallforms.

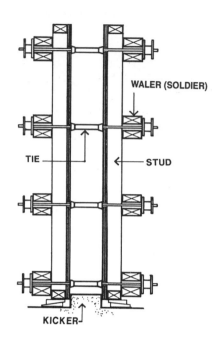

Figure 4.08 - HORIZONTAL WALER (SOLDIER) FORMWORK

Figures 4.09(a), (b) and (c) show the load path from the place of application of the fluid concrete pressures, the form face, to the studs, then the walers and finally to the ties. Firstly, the concrete pressure acts on the plywood of the formface. This plywood spans horizontally between the vertical studs and passes the loads to the studs. (Figure 4.09(a)).

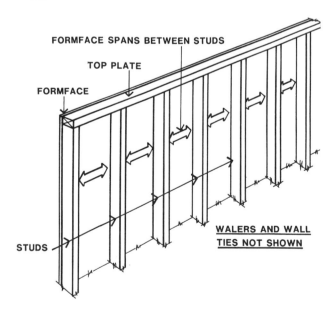

Figure 4.09(a) - PLYWOOD SPANNING BETWEEN STUDS

It is normal practice to fix the plywood sheets with their long dimension horizontally. As a result the plywood is acting as a flexural member which is continuous over a large number of spans. For a 2400 mm long sheet with studs at 200 mm centres, it would be continuous over 12 spans, a very efficient arrangement.

The studs span vertically between the walers and in turn pass their the loads to the horizontal walers. This section of the load path is illustrated in Figure 4.09(b).

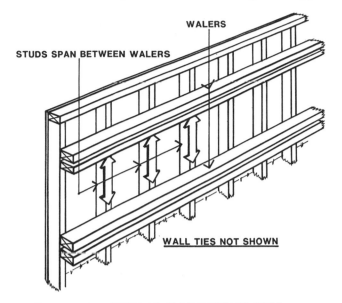

Figure 4.09(b) - STUDS SPAN BETWEEN WALERS

The horizontal walers then carry the loads to the ties (Figure 4.09(c)). As the other and opposing face of the formwork for the concrete wall has identical loads, the forces on the two form faces balance each other at the ties.

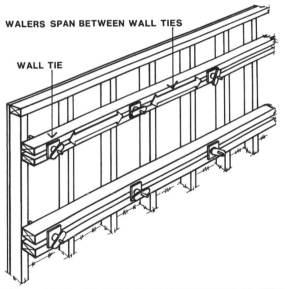

Figure 4.09(c) - WALERS SPAN BETWEEN WALL TIES

The walers are shown as twin members, but they could be single. However, there are several advantages in using twin members for walers. The use of twin members avoids drilling for the tie rods and the consequent loss of strength that occurs with single members. They also have advantages in the installation of the tie rods and alignment of adjacent formwork units. These matters are discussed later in this chapter.

Although the studs usually have two or more spans from the top waler to the bottom waler, it cannot be assumed that they always act as continuous beams with two or more spans. Not all of the wall tie systems are able to act to stop the inwards movement of the formwork framing. As the pour progresses up the wall there is a stage, as shown in Figure 4.10, where only the lowest stud

span is loaded. With the tops of the studs not restrained against inwards movement, the studs are not able to span continuously. They tend to act as simply supported beams. The design calculations should take account of this.

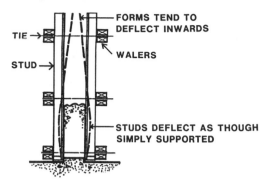

Figure 4.10 - SIMPLY SUPPORTED SPANNING ACTION

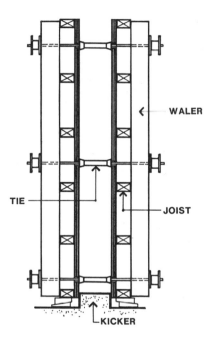

Figure 4.11 - VERTICAL WALER FORMWORK

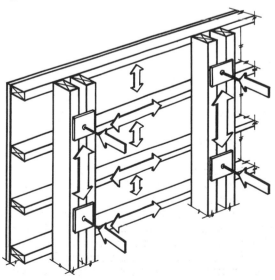

Figure 4.12 - LOAD PATHS FOR VERTICAL WALER FORMS

Figure 4.11 shows the cross section of a vertical waler formwork assembly. Here the plywood spans vertically to pass the concrete pressure loads onto the joists. These span horizontally to carry the loads to the vertical walers. In turn, the walers span vertically to carry the loads to the ties where the forces are balanced by the equal and opposing forces from the other formed face. Figure 4.12 shows this load path.

As stated in Chapter 2, and explained in more detail later in this chapter, the fixings of formface to framing and fixings between the framing members do not usually play any part in transmitting the forces along these load paths in either type of wall formwork.

Selection of Wall Formwork Type

Both of these wall formwork types are suitable for an isolated wall. However, where the walls to be built are especially big, long or tall, or the work is to be a sequence of adjoining pours, both types have their own particular advantages.

Where the length of the formwork unit is long or there is to be a linear sequence of walls, each as an extension of the previous pour, then the horizontal waler form can be best. If properly fixed to the studs, the walers can give the formwork the longitudinal stiffness needed for a long form to be safely hoisted.

When a wall to be formed is an horizontal extension of a previous pour then the walers can be extended back along the face of the previous pour and used to align the work, the new with the older.

The length of the formwork need not be limited by the available length of waler timbers. The use of double members for walers permits them to be effectively lapped without loss of their ability to control alignment. Figure 4.13 illustrates this.

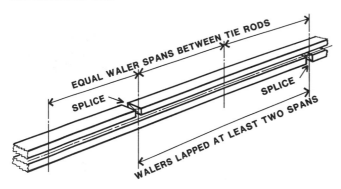

Figure 4.13 - LAPPING HORIZONTAL WALERS

For similar reasons vertical waler formwork has advantages for tall wall forms, and those where subsequent pours are to be a vertical extension of the previous pour. When the formwork is stripped after the first use and hoisted up to its next position, the extended vertical walers are rearranged to lap down over the wall below.

The tie positions from the earlier work are reused to clamp the formwork. The tie positions at the bottom of the walers have either packers or screw jacks fitted. These are used to control the plumb of the formwork.

If the forms are heavy, a special support bracket can be incorporated in either the upper or the lower ties. Figure 4.14 shows a simple example of this climbing situation.

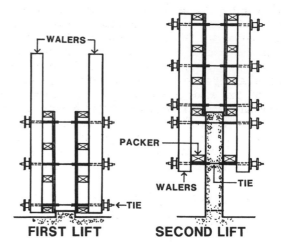

Figure 4.14 - CLIMBING FORMWORK

Deflection of Framing Members

When they come under load, all flexural members in the load path deflect, even if it is only a very small amount. While it is not the intent of this book to cover the calculation of these deflections, an understanding of their effects is essential to an appreciation of how efficient formwork should be built.

Specifications for the concrete work and the associated codes and standards give the tolerances for acceptable work, that is, the final concrete shape that will be judged as meeting the requirements of the specification. Tolerances are defined as the maximum permitted deviations from perfection.

These deviations are caused by formwork fabrication errors and deformation of the formwork and its supports under load. The deformations comprise deflections of flexural members and the 'take up' of the components under load. For walls, examples of 'take up' are timber crushing under washers to tie rods and between walers and studs where they intersect.

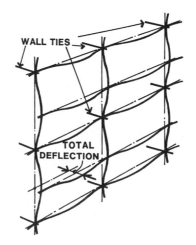

Figure 4.15 - CUMULATIVE DEFLECTIONS

In assessing the effect of deflections, both on the total deviations and the visual quality of the surface, it must be appreciated that they can be cumulative. Figure 4.15 indicates the deflections for a vertical waler form. The total deflection for the horizontal joist at mid-span of the walers is the sum of their individual deflections. To this can be added any fabrication errors and 'take up'.

Another practical effect of deflection can be grout loss. Refering back to the cross section of Horizontal Waler Formwork, Figure 4.08; a bottom plate is shown at the base of the studs even though the plywood spans horizontally between the studs. That is, it spans parallel to this bottom plate. The intent of this bottom plate is the elimination of deflection of the plywood at its lap onto the kicker. Without the bottom plate, deflection of the plywood along the line of the kicker would result in considerable grout loss.

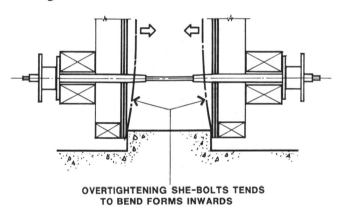

Grout Loss at Base of the Formwork

To increase the effectiveness of the bottom plate, and the clamping of the forms onto the kicker, the walers are also lapped onto the bottom plate. This feature was shown in Figure 4.08.

OVERTIGHTENING SHE-BOLTS TENDS TO BEND FORMS INWARDS

Figure 4.16 - INCORRECT BOTTOM WALER LOCATION

If the walers are located above this bottom plate, even if only a short distance, the tightening of the ties to clamp the forms to the kicker will result in bending of the forms (Figure 4.16). Even a small amount of flexure in the formfaces will produce a noticeable misalignment in the final concrete faces. The curved shape of the forms also moves the point of application of the clamping action to the top corner of the kicker. Crushing of the formface and local fracturing of the edge of the kicker can result.

Where the framing members are cold rolled metal channels this bottom plate can be installed as shown in Figure 4.17. The bottom plate is fixed to the studs with bolted angle cleats, and the double walers are clipped to the studs. All these components are standard industrial accessories produced, like the 'C' section framing members, for purlins and girts for industrial sheds.

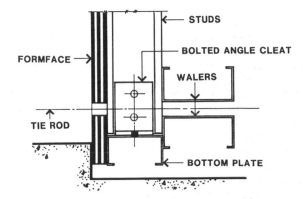

Figure 4.17 - BOTTOM OF METAL FRAMED FORMS

As can be seen from Figure 4.11 Vertical Waler forms already have a stiff framing member to limit grout loss at the kicker and no additional framing is required. However, both types of formwork require special attention to prevent grout loss at construction joints and stop ends. Both of these topics are dealt with in detail, later in this chapter.

WALL TYING SYSTEMS

The function of formwork tying systems is to connect the two faces of the double faced forms together and effect the balance of the forces between them. The spacing of the ties, and the forces they have to resist, is determined by the wall to be built and the framing system adopted.

Wall ties, also known as tie rods, tie bars and tie bolts, can be described under three broad groupings: He-bolts, She-bolts and Through ties. The main types available will described firstly then their installation procedures will be discussed.

He-bolts

He-bolts are so named because the connection to the formwork framing is with a male threaded bolt. The simplest example, shown in Figure 4.18, is used with a type of all-metal proprietary formwork which is described later in this chapter.

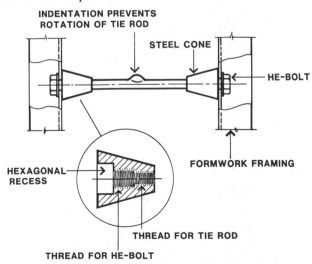

Figure 4.18 - HE-BOLT FOR PROPRIETARY FORMWORK

The male threaded bolts connect, through the formwork framing, to steel cones. These connect across the wall with a high-tensile tie rod. Tie rods are available in a variety of lengths to suit common wall thicknesses. The tie assembly not only resists the tensile forces from the forms to effectively hold the wall forms together, but the shoulders of the cones also hold the form faces the correct distance apart at all times.

Removal of the bolts enables stripping of the forms. The cones are recovered by screwing them off the tie rod with an Allen key inserted in the hexagonal recess on their outer face. The tie rod remains in the wall and is provided with an indentation or kink to prevent it rotating when the cones are being screwed out.

The removal of the cones, or at least their loosening, should be done soon after the forms are stripped. To delay this may result in the cones being very difficult to remove. This will be due to drying shrinkage of the concrete. Later, the cone holes are either mortar packed or plugged with pre-made cones bonded into place.

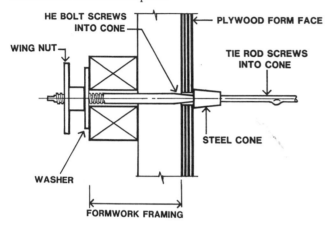

Figure 4.19 - HE-BOLT WITH TIMBER FORMS

Figure 4.19 shows the He-bolt style adapted for conventional timber framed formwork systems. The He-bolt is extended to have a larger shaft with a threaded section for a large wingnut and washer. The outer end of the bolt has two flats so that a spanner can be used to screw it out of the cone. This unit effectively clamps the formwork framing to the plywood and the tight interface between the cone and the plywood face can prevent grout loss at the formface.

However, care should be taken in tightening the wingnut to achieve this clamping action. Two problems can occur from overtightening the wingnut. Firstly, the cone can crush the surface veneer of the plywood and thus reduce its working life.

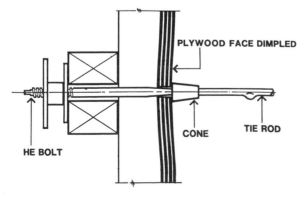

Figure 4.20 - OVER-TIGHTENING OF THE HE-BOLT

Secondly, when the tie is located approximately midway between the studs of the formwork framing, the overtightening can pull the plywood outwards with a resultant pronounced dimpling of the concrete face. (Figure 4.20).

As noted previously, ties from a previously poured concrete wall are often used to clamp the formwork on for the next pour. These forms are, of course, erected one at a time. Because the separate steel cones effectively lock the tie rod into the hardened concrete, the He-bolt types can clamp a new formwork wall unit on, one side at a time. Figure 4.04 showed this climbing form situation. Most of the other types of ties do not have this facility and must be completely assembled with washers and wingnuts at both ends to give effective anchorage.

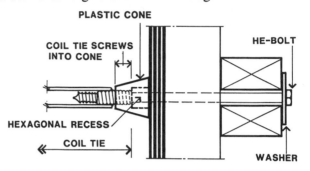

Figure 4.21 - COIL TIE

Coil ties, shown in Figure 4.21, are another version of the He-bolt. The tie, which remains in the concrete, consists of two side rods which are welded to tightly wound helical coils of thick wire. The bolts (He-bolts) thread into these coils, and have the sharp pointed thread profile shown to fit the spaces between the wires.

Plastic cones screw over the outside of the coils at their ends. The cones provide a good seal against grout loss at the formface. They are available in a variety of lengths (colour coded) to suit the specified reinforcement cover and give overall tie length adjustment. When the concrete is placed it encases the end of the bolt, and grout usually penetrates between the bolt and the inside of the coil. Two matters arise from this.

Firstly, the concrete encasement of the bolt means that as it is being screwed out there is abrasion and wear of the thread. After a number of uses the sharp tips of the threads can be well worn. This weakens the bolt and increases the chance of failure under load. At each use, bolt for coil ties must be carefully examined for wear.

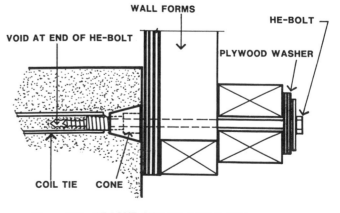

Figure 4.22 - SECOND USE OF A COIL TIE

The second problem occurs if a coil tie from a previous pour is to be used for fixing the forms for the next adjacent pour. The concrete encasement of the end of the bolt, at its first use, limits the distance that the bolt can be screwed back into the coil tie. If the thickness of the framing to the new formwork is slightly less than that of the forms to the earlier pour then the bolt cannot effectively clamp the forms to the concrete face. As Figure 4.22 shows, the insertion of a thick washer, say 12mm plywood, at the bolt head prevents its full insertion and the formwork can then be tightly clamped to the concrete face.

She-bolts

As the name implies, the connection to the tie rod is by female thread. Figure 4.23 shows the She-bolt screwed onto the end of the tie rod. These tie rods are the same as those used with He-bolts.

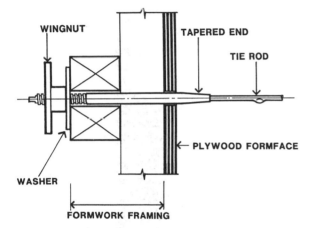

Figure 4.23 - SHE-BOLT WITH TIMBER FORMS

The end of the She-bolt, at the connection to the tie rod, is tapered to enable its removal after the pour. The other end is threaded for a large wingnut and washer, similar to a He-bolt. The She-bolt has no interface at the plywood surface to seal against grout loss. Consequently grout leaks along the shaft of the bolt, even if the hole in the plywood is a tight fit.

She-bolts only provide a tensile capacity to resist the forces from the concrete pressures on the forms. They do not hold the forms apart or in any way contribute to their alignment or the control of wall thickness. If control of the wall width is needed other means must be used.

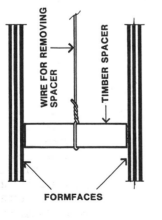

Figure 4.24 - REMOVABLE TIMBER SPACER

For short wall forms, the simplest control of wall width is the nailing of battens across the top of the two forms. For taller forms some internal type of control of wall width is needed.

For control of wall width within the form Figures 4.24 and 4.25 show two methods. The first (Figure 4.24) involves retrievable spacers. Pieces of battens, usually 50 x 25 mm in section, are placed between the form faces. The tightening of the She-bolts holds them in place. As the level of the concrete placing rises towards them, they are pulled out using the wires shown. Unfortunately, they are often forgotten and have to be dug out of the hardened concrete after the forms are stripped.

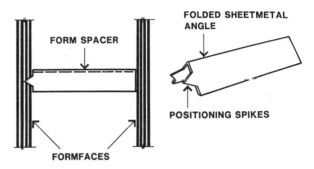

Figure 4.25 - PERMANENT WALL SPACERS

Alternatively, permanent wall spacers can used (Figure 4.25). These proprietary folded sheet metal angles spike into one form face and the tightening of the wall ties secures their position. Care must be taken after stripping as the spikes protrude from the wall face and are a safety hazard. The spikes should be bent over or ground off.

Despite these shortcomings She-bolts are quite popular in the formwork industry. As later discusssions will show, they are easy, perhaps the easiest, to install and are available in a wide range of sizes (and strengths).

Through Ties

Through ties are so named because the tensile loads are carried on one single member that passes through the formwork-wall-formwork assembly from end fixing to end fixing. The oldest of these, and still occasionally used these days for in-ground work, is the twisted wire loop. Figure 4.26 shows an example of its use.

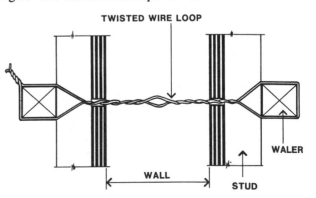

Figure 4.26 - TWISTED WIRE LOOP WALL TIE

Other long established through ties, that are still occasionally used despite their low load capacity, are those comprising a plain round rod as the tension member, with end anchors devices. To ease removal of the rod, a plastic tube, with its length cut to the wall thickness, can be placed over the rod. Figure 4.27 shows one these where the rod is held by a bolt tapped into the anchor plate. Because the bolt kinks the rod sideways, it is more effective than the conventional grub-screw. By nailing the anchor plate to the formwork framing this type of tie can also give some control over the wall thickness.

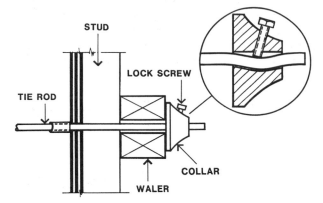

Figure 4.27 - STEEL ROD AS WALL TIE - 1

In the second example, shown in Figure 4.28, the rod is anchored with a grooved wedge. As the wedge is driven in, the multiple grooves deform the outside of the rod. This surface deformation of the rod greatly increases the anchoring action created by the wedge.

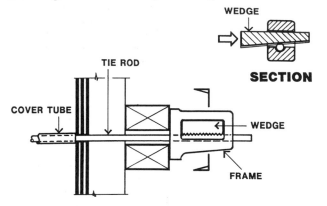

Figure 4.28 - STEEL ROD AS WALL TIE - 2

Snap-ties are a type of through tie which enable the ends of the ties, that protrude from the concrete face, to be removed by breaking them off. These ties are provided with a weak point, a short distance in from the concrete faces, at which the break can be effected. Figure 4.29 shows three examples of Snap-ties.

As indicated on the diagrams, this snapping action also breaks some of the concrete away. The resulting jagged areas must be patched. This also makes this type of tie quite unsuitable for the production of high quality concrete surfaces.

Examples numbered 1 and 2 are used with proprietary modular formwork systems. These systems are discussed later in this chapter. Example 1 is fabricated from high tensile rod and a flattened, and weakened, section is formed just within the concrete. Twisting the tie breaks it. At the same time some of the concrete is broken out. Example 2 made from a steel flat is weakened by

notching. Bending the tie downwards, usually by striking with a hammer, breaks the tie.

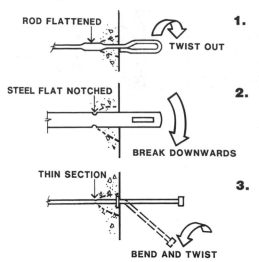

Figure 4.29 - THREE TYPES OF SNAP-TIES

The third example is made from a steel rod with an enlarged head formed on each end and thin sections located at the required break point. They are used with conventional framed forms as shown in Figure 4.30. The ties are usually fitted with a metal washer located at the formface. Their tightening action is achieved with a slotted steel wedge that fits over the enlarged ends of the tie. Therefore, these ties not only provide tensile capacity but also control the wall thickness and clamp the formwork framing and plywood together.

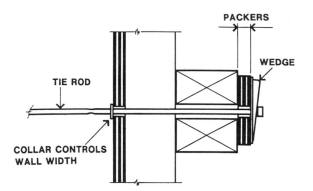

Figure 4.30 - SNAP-TIE WITH TIGHTENING WEDGE

These ties come in a range of sizes with the washers located to suit common wall thicknesses, and the distance between the washers and the ends sized to suit typical framing dimensions. If the framing is thinner than this, each wedge can be packed with an appropriate piece of plywood. Figure 4.30 illustrates this.

The through tie shown in Figure 4.30 has quite a low tensile capacity. In contrast, the type of through tie illustrated in Figure 4.31, the bar tie, has the highest tensile capacity of the tying systems currently available.

The tie is a threaded high tensile rod which is cut from a stock length to suit the particular use. The thread is of the rounded "rope-thread" type which is not readily prone to wear. As Figure 4.31 shows the complete assembly consists of a plastic tube fitted with plastic cones at its ends, the bar tie passing through the tube with washers and wing nuts at the ends.

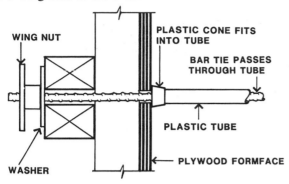

WING NUT

PLASTIC CONE FITS INTO TUBE

BAR TIE PASSES THROUGH TUBE

PLASTIC TUBE

PLYWOOD FORMFACE

WASHER

Figure 4.31 - BAR TIE WITH TIMBER FORMS

The tube and cone assembly holds the formfaces apart the correct distance, and the tie bar and wingnuts provide the needed tensile capacity and thereby clamp the whole assembly together.

However, because there is no connection between the bar tie and the tube assembly, there is no individual clamping action on the formwork assemblies at each side. When used with a climbing formwork procedure, it cannot be relied on to clamp one form to the previously poured concrete wall like a he-bolt can. (refer to Figure 4.04)

In this respect it is different to the he-bolt, but like the he-bolt and coil tie, the cones give a grout tight junction at the plywood faces. These cones and the tube, being plastic, can only be compressed a limited amount and this has the advantage of minimising any overtightening and the resultant dimpling of the formface.

The bar tie with its continuously threaded rod, wing nuts and washers, is useful in other formwork applications where a tensile tie is needed. One example is external ties for column forms. (Chapter 5)

Ties for wall forms are selected on the basis of required tie capacity, the need, if any, to eliminate grout loss at the tie holes, and the ease of installation. The installation of he-bolts, she-bolts and bar ties is shown below.

Installation of Wall Ties.

The particular characteristics of each of the types of wall ties calls for a sequence of installation to suit each case. The illustrations of these procedures, given below, omit to show any bracing to the forms in order to simplify the diagrams. However, for reasons of safety, it is needed at all times.

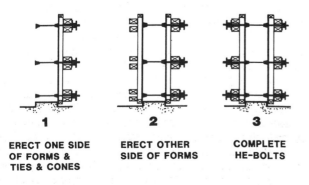

1
ERECT ONE SIDE OF FORMS & TIES & CONES

2
ERECT OTHER SIDE OF FORMS

3
COMPLETE HE-BOLTS

Figure 4.32 - INSTALLATION OF HE-BOLTS

Figure 4.32 shows the installation sequence for he-bolts and coil-ties. One face of formwork is erected, the tie rods and cones are installed on this face with one set of he-bolts, washers and wingnuts. After reinforcement tying, the opposite formface is erected and the other ends of the he-bolt assemblies can then be installed.

This last operation is not always straightforward, and some juggling of the he-bolts may be needed to successfully insert them into the cones. Positioning the walers to aid this operation is discussed under 'Typical Wall Formwork Fabrication and Erection'. For coil-ties the operation is usually a little easier than for He-bolts due the design of the plastic cones and the pointed ends of the bolts.

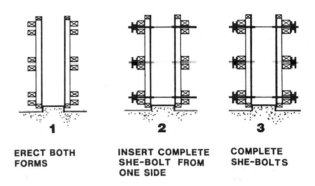

1
ERECT BOTH FORMS

2
INSERT COMPLETE SHE-BOLT FROM ONE SIDE

3
COMPLETE SHE-BOLTS

Figure 4.33 - INSTALLATION OF SHE-BOLTS

She-bolts can be installed in the same way as he-bolts. However, the absence of a cone at the formface makes a simpler sequence possible. (Figure 4.33) Both formfaces and reinforcement can be erected first with care being taken to keep the reinforcement clear of the she-bolt positions.

With the wing nut and washer removed from one end, the whole she-bolt assembly can be inserted from one side. Then these washers and wing nuts are fitted to complete the wall ties.

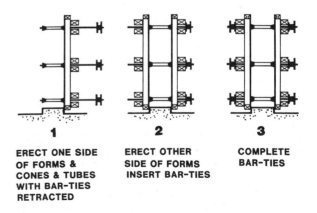

1
ERECT ONE SIDE OF FORMS & CONES & TUBES WITH BAR-TIES RETRACTED

2
ERECT OTHER SIDE OF FORMS INSERT BAR-TIES

3
COMPLETE BAR-TIES

Figure 4.34 - INSTALLATION OF BAR-TIES

Bar ties are installed similarly to he-bolts. However, the lack of any positive connections between the bar and the tube/cones makes the procedure more difficult. After the erection of one formface and the reinforcement, the bar ties are inserted through the formface with the cones and tube placed loosely over the protruding ends. (Fig. 4.34) The bars are retracted so that they project only a short distance beyond the cone.

The other form is then erected in a position a short distance from the ends of the bars. After the ends of all the bar ties have been inserted in the holes in the form, this form is moved up to its correct position and the assembly of the ties completed.

All the ties and the associated assembly procedures that have been mentioned above are related to double sided forms. At the beginning of this chapter mention was made of single faced forms. As will be shown later these formwork systems sometimes use the wall ties described above, connected to anchors installed in previously poured concrete. Special tie anchoring systems are used for this. Figure 4.35 shows the pig-tail anchor used with she-bolts or he-bolts. Figure 4.36 shows the coil-tie loop anchor which has the same function, an anchor for later use.

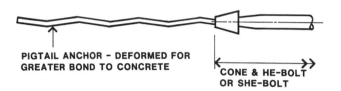

PIGTAIL ANCHOR - DEFORMED FOR GREATER BOND TO CONCRETE

CONE & HE-BOLT OR SHE-BOLT

Figure 4.35 - PIG-TAIL ANCHOR

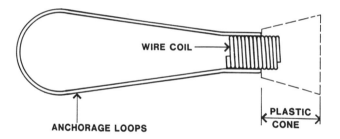

WIRE COIL

ANCHORAGE LOOPS

PLASTIC CONE

Figure 4.36 - COIL TIE LOOP ANCHOR

Figure 4.37 illustrates another type, the anchor screw. The hexagonal headed bolt and the helical thread are cast into the concrete. The bolt head protrudes beyond the concrete face by approximately 5 mm less than the grip needed for the assembly, formwork or fitting, which is to be bolted later to the hardened concrete. This will enable effective tightening without the bolt hitting against the end of the hole in the concrete.

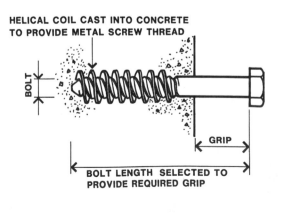

HELICAL COIL CAST INTO CONCRETE TO PROVIDE METAL SCREW THREAD

BOLT

GRIP

BOLT LENGTH SELECTED TO PROVIDE REQUIRED GRIP

Figure 4.37 - ANCHOR SCREW

TYPICAL WALL FORMWORK FABRICATION AND ERECTION.

One of the principal objectives of the various assembly techniques used in formwork fabrication and erection, is the production of a strong rigid accurate form that can be easily handled in erection, alignment and, finally, stripping. A form must repeatedly withstand the vigorous actions of formworkers, reinforcement fixers and concretors. The achievement of this aim starts with the fixing of the formface.

Formface fixings must keep the joints tight, fix the plywood to the framing so that the formwork assembly is a cohesive structure, and enable ease of dismantling resulting in maximum material recovery.

The load path on wall formwork during concrete placement was described earlier in this Chapter. From this it can be seen that formface fixings such as nails and screws are not part of this load path; they do not normally serve a purpose in this structural action. The fixings holding the plywood to the studs, and the studs to the walers, play no part in the resistance to the concrete pressures that act on the formface.

Their purpose is to hold the formwork unit together while it is being made, erected and climbed on by workers. Except for small forms, minor fixings like nails do not usually give the form sufficient strength or rigidity to resist hoisting by crane. Where crane hoisting or a large number of reuses is expected, attention must be given to the type and spacing of the plywood fixing. This topic is covered in more detail later in this chapter.

For a horizontal waler wall formwork assembly, a typical fabrication sequence would start with the assembly of the studs and top and bottom plates. Uniformity of depth of studs and plates is essential to get a true formface. As discussed in Chapter 2, this can be achieved by using 'sized' timbers.

The fixing used to connect these members to each other with depend on the size of the form and the way it is intended to handle and hoist it. At the least, they must be securely nailed together.

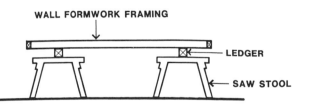

WALL FORMWORK FRAMING

LEDGER

SAW STOOL

Figure 4.38 - ASSEMBLY OF FORMFACE TO STUDS

Figure 4.38 shows this assembly being done on a workbed of two heavy ledgers mounted on saw stools. To avoid building a twist into the form, the ledgers should be set level and parallel. The walers can be used for these ledgers but they should not be fixed to the studs. For accuracy, this is best done after the tie holes have been drilled. As an alternative to the ledgers, an accurate concrete floor serves equally well.

In addition to load/span considerations for the selected plywood, the spacing of the studs must suit the sizes of the plywood sheets being used and the intended position of the junctions of the plywood sheets.

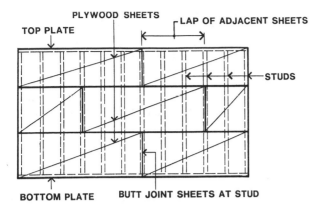

Figure 4.39 - LAPPING OF PLYWOOD SHEETS

The pattern of plywood sheets shown in Figure 4.39 greatly aids the overall stiffness of the form. Lapping the plywood sheets in this way gives some effective structural continuity, and nailing produces some composite structural action of the plywood and the framing.

Care should be taken during nailing (Figure 4.40) to ensure that the carpenters weight does not cause deflections in the framing which will become built into the completed form by this composite action. Such misalignments can be nearly impossible to correct without at least partial dismantling of the form.

Figure 4.40 - FORMWORKER CAN CAUSE DEFLECTION

The best time to drill the holes for the tie rods is when the fixing of the plywood to both forms is complete, but before the access platforms and walers have been added. Their position is controlled by a number of construction considerations as well as any visual pattern requirements which may have been specified.

It is common practice for specifications for high quality concrete surfaces to call for strict control of the regularity of the pattern of tie rod holes, and submission of details for approval. In some cases the positions of the ties may be dictated.

Although the structural design of the wall form gives the maximum spacing of the ties, other practical considerations must also be taken into account when positioning the rows of ties. The need for the bottom line of walers to stiffen the bottom plate of the wall was illustrated in Figure 4.08. The problem that came from having it too high was shown in Figure 4.16.

The location of the top walers may well be influenced by the need to use its ties for another and later formwork construction. This may be for the support of the next lift of wall forms or as a fixing line for a bearer or bearers to

support formwork for the soffit of a slab. Figure 4.41 shows an example of this latter case.

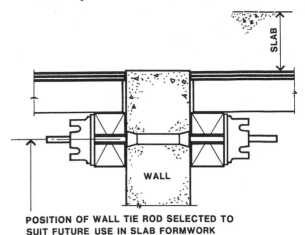

POSITION OF WALL TIE ROD SELECTED TO SUIT FUTURE USE IN SLAB FORMWORK

Figure 4.41 - USING THE TOP TIE ROD HOLES

To ensure a perfect match of tie rod holes in the opposing forms, they should be drilled together. The two forms should be placed face to face. The tie holes are then set out and drilled through both formfaces as shown in Figure 4.42.

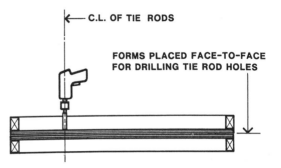

Figure 4.42 - DRILLING THE FORMS FACE TO FACE

To place both the forms the same way up, or back to back, risks the misalignments of holes from inaccurate drilling. (Figure 4.43)

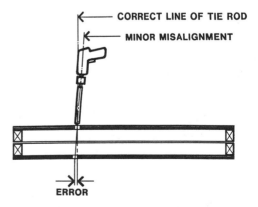

Figure 4.43 - ERROR FROM DRILLING BACK TO BACK

Alternatively, the holes can be accuratively set out on each formface and then drilled. However, with this method, the possibility of a mismatch always exists.

The walers can now be added. The hole positions give the line of the centre line of the waler pairs. The distance between the pair of walers should be at least 10 mm more than the diameter of the wall tie. This enables easier installation of the tie. It can be moved about for alignment with the holes in the opposite formface. (Figure 4.44)

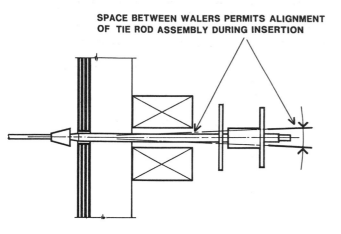

SPACE BETWEEN WALERS PERMITS ALIGNMENT OF TIE ROD ASSEMBLY DURING INSERTION

Figure 4.44 - EASE OF TIE ROD ALIGNMENT

At the least, the waler fixings to the studs will be skew nailing. For large forms and those cases where the walers are part of the hoisting method, the fixings may be screws or bolted angle cleats.

It should be noted that the walers are not always added at this stage. For small forms the walers are sometimes individually put in place above and below the ties after the ties have been installed.

After the walers, any access platforms and lifting point hardware should be added. The details of the construction of these is covered later in this chapter. The form faces are ready for hoisting.

One face of the forms is now hoisted into position and braced. (Figure 4.45) This bracing may be temporary, with the final bracing system being completed later. To prevent rotation of the form under wind load the bottom as well as the top should be braced. (Refer to Figure 4.03.) The reinforcement is now tied, stopends, penetrations and built-in fittings fixed in place and the installation of the ties started (If the tie rods are She-bolts they can be installed later).

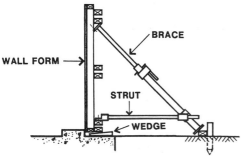

Figure 4.45 - TYPICAL WALL FORM BRACING

The other side of the formwork can now be hoisted, the tie installation completed, the stop-ends and final bracing completed and the formwork's line, level and plumb adjusted. The line of the bottom is fixed by the

kicker. The line of the top and the plumb of the wall is corrected by adjusting the bracing. The levels of the forms are usually adjusted with wedges at their base.

BRACING WALL FORMS

The loads on wall forms that the bracing has to resist were discussed earlier in this Chapter. Firstly, there is wind, acting at all times and from any direction. Secondly, during construction, while either building the formwork or pouring the concrete, there is the possibility of impact. This is usually from crane hoisted loads.

To brace wall forms with adjustable telescopic props can be satisfactory as a temporary measure, but they will be potentially dangerous unless positively fixed in place . Even a light blow will knock them out of place and the wall form may topple over.

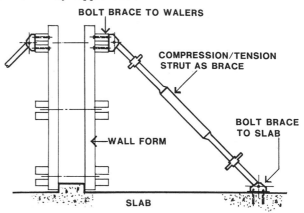

Figure 4.46 - DOUBLE ACTING PROP BOLTED IN PLACE

It is essential that it should not be possible for the bracing to be dislodged by impact. If impact occurs the bracing can be badly bent, but, as long as it is still attached to its connection then it will prevent catastrophic collapse and the generation of dangerous debris. One solution is shown in Figure 4.46. Bolting to the walers and the slab means that bending of the brace from impact will be unlikely to totally remove the brace. However, this method is costly and time consuming. Figure 4.47 shows another method.

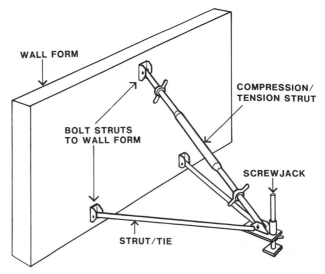

Figure 4.47 - BRACING FRAME

This involves the bolting of sets of bracing frames to both sides of the wall formwork assembly. Most wall forms would require at least two sets on each side, long forms would need more. Each set comprises a raking strut up to the top of the wall form and two low level struts, all connected together and to the wall form. This triangular framing can give resistance to impact from all directions.

Even if impact badly distorts the frame, it will still give some bracing effect provided the fixings do not fail. Sliding of the whole form assembly will be prevented by the wall kicker or the wall reinforcement. The important point is that impact should not be able to totally remove the bracing.

DETAILS OF CONSTRUCTION

Plywood Fixing to Framing.

As noted previously, the primary function of the plywood fixings is the production of a form that can withstand the activities of handling, erection, alignment and stripping. The fixings hold the formwork unit together while it is being made, erected and climbed on by workers.

Beyond that, the plywood fixings may also need to be sufficient to make the plywood act in conjunction with the framing to give the strength and rigidity needed for crane hoisting. For this, nails are usually inadequate; except for cases of small wall forms, screw fixings are essential.

For nailing, the traditional formula is for the nail to be 2.5 times the plywood thickness in length. For formwork a slightly longer nail is recommended: 50 mm long and at least 2.5 mm in diameter for 17 mm plywood. Flatheaded nails are the most commonly used type for general formwork. The large head does not readily pull through the plywood when the form is being knocked apart. It also provides a good gripping force as a fixing of plywood to framing.

While flatheads are quite good for general formwork, they leave an unsightly mark on the concrete face which is usually not acceptable for high quality off-form surfaces. If nailing is permitted, it is best done in this case with bullet head nails, despite their tendency to pull through during formwork dismantling. The recommended size is also 50 x 2.5 for 17 mm ply.

To disguise the nail positions, the heads are punched down, stopped flush and then lightly sanded. The stopping must have a very low moisture absorption to avoid the nail positions showing up as small areas of hydration staining.

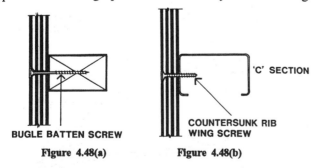

Figure 4.48(a)   **Figure 4.48(b)**

SCREW FIXINGS THROUGH THE FORMFACE

Screws through the face, if surface finish requirements permit, should be 'Bugle Batten' type No. 10, 50 mm long

for fixing 17 mm plywood to timber framing. (Figure 4.48(a)) If the framing consists of cold formed metal framing members, e.g. Zed or 'C' sections, countersunk screws of the 'Rib Wing' type are recommended. (Figure 4.48(b))

For quality work, screw fixings from the back can give the best results. To fasten effectively, the screws must penetrate at least the plywood thickness less 3mm. They must not break through the formface. If a standard length screw will not achieve this, then a longer one can be used with washers under its head to reduce its effective length.

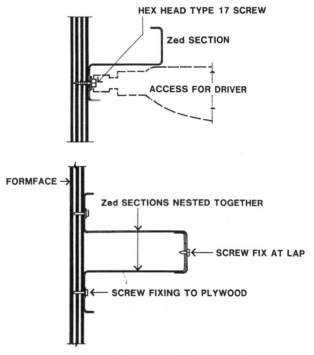

Figure 4.49(a) - BACK FIXING Zed SECTIONS

Figures 4.49(a) to (d) show four examples of screw fixings from the back face of the plywood. For rear fixing, Zed sections are the most convenient cold rolled framing due to the ease of access for the power screw driver. (Figure 4.49(a)) For these thin metal members the 'Type 17' screws are self drilling. For thicker sections, predrilling is needed. (Figure 4.49(b))

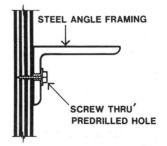

Figure 4.49(b) - BACK FIXING PREDRILLED ANGLES

For timber framing, rear screw fixing direct from timber section to plywood requires consistently accurate workmanship. As Figure 4.49(c) shows, a small deviation in the angle at which the screw is driven can cause either formface break through or insufficient penetration of the plywood.

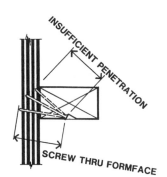

Figure 4.49(c) - BACK FIXING TIMBER FRAMING

A better method, for use with timber framing, is the screw fixing of a small steel angle to the side of the timber framing members. (Figure 4.49(d)) Screws are then driven through predrilled holes in the angle into the back face of the plywood. Setting the angle with a small gap between it and the plywood, helps screw tightening and can also be used for adjustment to suit the available screw length.

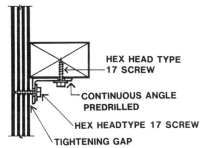

Figure 4.49(d) - ANGLE FIXING FOR TIMBER FRAMING

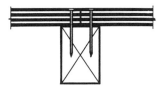

Figure 4.49(e) - JUNCTION OF PLYWOOD SHEETS

Where plywood sheets butt, the width of the framing member must be wide enough for two rows of fixings. (Figure 4.49(e)) A width of 50 mm is a minimum, 75 mm is ideal. A similar requirement can occur at the junction between forms.

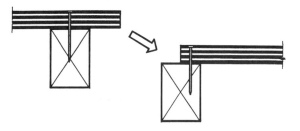

Figure 4.50 - JUNCTION BETWEEN TWO FORMS

If the total formwork arrangement, such as a long wall form, is made up from the site assembly of a number of shorter forms, alignment of the formfaces at their junctions

is important. One suitable method is shown in Figure 4.50. The plywood projects from one form, to seat on half of the face of the edge framing member of the adjacent form. A member width of 50 mm is the minimum to give a 25 mm bearing width, 75 mm is better.

In both these examples of the junctions of plywood sheets, a foam plastic filler tape, placed between the ends of the plywood sheets, can be used to limit moisture loss through the joint and into the ends of the plywood sheets.

Framework Fixings.

Reference was made earlier to the minimum fixings between framing members being secure nailing. As an assembly procedure this is almost always used. Other, stronger, fixings may need to be added to suit the selected means of handling and hoisting the formwork.

Some information on these matters will be given in Provision for Hoisting Wall Forms' later in this chapter.

Stop-Ends and Construction Joints

At the ends of walls, and at the planned positions of construction joints, stop-ends must be constructed. Due to reflection and rebound of the energy of vibration from the face of the stop-end, the concrete pressure acting on it can exceed that generally acting on the formface. (Figure 4.51) Hence, special attention must be given to preventing grout and moisture loss and to the fixing of the stop-end.

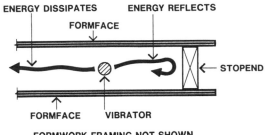

Figure 4.51 - REFLECTION OF VIBRATOR ENERGY

In most cases, the studs of the formwork do not coincide with the stop-end and, as a result, the plywood deflects under the concrete pressure and grout and water escapes past the stop-end. (Figure 4.52) At the least, there is hydration staining due to moisture loss. Often it causes severe honeycombing.

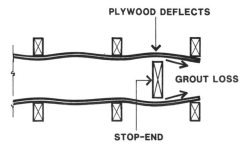

Figure 4.52 - FORMFACE DEFLECTION AT STOP-END

This plywood deflection, although quite acceptable within the general area of the form, must be prevented at the stop-end. For horizontal waler forms, an extra packer stud is placed on the line of the stop-end on both faces. This is shown in many of the details that follow.

For vertical waler forms, short vertical packers are placed between the joists and beside the stopend. Ideally, a vertical waler should coincide with them. (Figure 4.53)

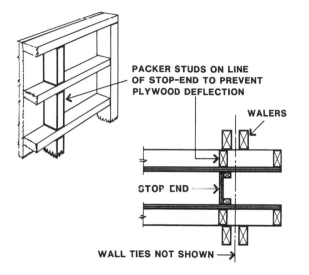

Figure 4.53 - PACKERS BESIDE THE STOP-END LOCATION

A common method of resisting the forces on the stop-end is by wedging it off the wall ties.(Figure 4.54) In addition, there is considerable friction created between stop-end and form faces with the tightening of the wall ties.

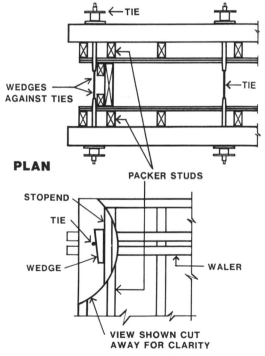

Figure 4.54 - WEDGING STOP-END OFF WALL TIES

Plywood bearing strips can also be used to carry the loads on the stop-end to the ties. If bar ties are used they can be fitted with extra wingnuts to clamp the plywood bearing strips to the form faces as shown in Figure 4.55.

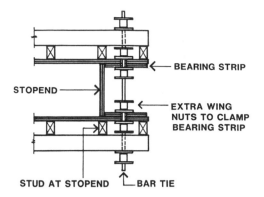

Figure 4.55 - BEARING STRIPS CLAMPED IN PLACE

In this case the stop-end is shown made of plywood spanning between the bearing strips. This is usually quite satisfactory for walls up to 250 mm thick. For thicker walls the stop-end may need have framing to the plywood.

If the wall ties are not so conveniently located just outside the stop-end, props can often be used to carry the loads on the stop-end. In Figure 4.56, the props are shown wedged against the starters bars of the next adjacent pour.

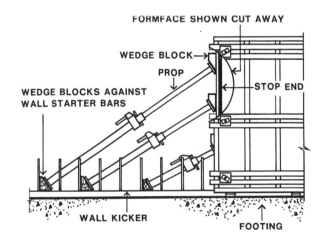

Figure 4.56 - STOP-END PROPPED OFF REINFORCEMENT

Alternatively, the stop-end can be tied back to the wall ties within the form. Figure 4.57 shows bar ties fitted with a popular brand of bar tie nut which is used as a hook. This links the bars ties to the wall tie. A pair of short walers bridges between the bar ties to hold the stop-end in place.

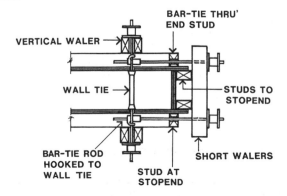

Figure 4.57 - STOP-END HOOKED TO WALL TIES

Stop-ends are not always just plane faces. Figure 4.58 shows the installation of a waterstop. Two shaped timber sections clamp the waterstop in place and they are backed up by another solid section.

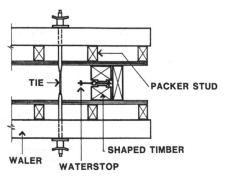

Figure 4.58 - STOP-END FOR WATERSTOP

At construction joints, stop-ends usually have to accomodate reinforcement. To aid stripping, the stop-end can be made in pieces as shown in Figure 4.59. This component is fabricated by drilling lines of holes in a single piece of timber and then sawing it lengthwise. Small cleats are used to hold the parts together.

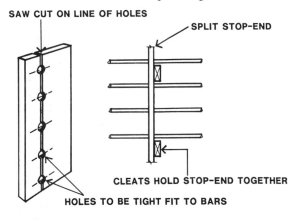

Figure 4.59 - CONSTRUCTION JOINT STOP-END

Another method involves using fine expanded metal mesh for the stop-end. Some loss of fines and moisture occurs but this has been found to be acceptable for general formwork use. These proprietary mesh stop-ends are usually produced with ribs to give them the strength to span across the width of the wall. Reinforcement can readily penetrate the mesh. In most cases its removal is not required for the pouring of the next section. For this reason, this mesh is kept back from the formface a distance equal to the reinforcement cover. (Figure 4.60)

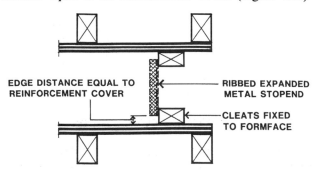

Figure 4.60 - PROPRIETARY MESH STOP-ENDS

It is not usual to install an extra stud to prevent plywood deflection in this case. The open nature of the mesh results in reduced rather than increased pressure at the stop-end.

It should be noted that the clamping of the forms to a previously poured concrete wall, to form the next section. It requires the same attention to plywood deflection that is needed for stop-ends. Packer studs are needed to stop the grout and moisture loss. (Figures 4.61).

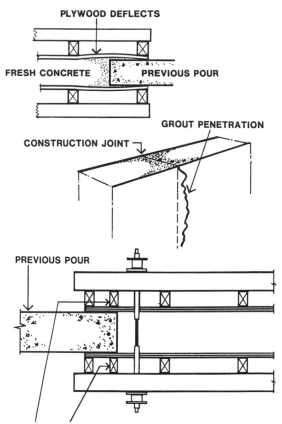

Figures 4.61 - JUNCTION OF FORMS TO CONCRETE WALL

Construction joints are sometimes needed at corners of walls and a horizontal kicker is useful for formwork alignment. (Figure 4.62)

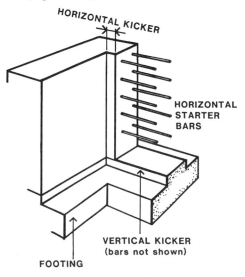

Figure 4.62 - HORIZONTAL KICKER FOR WALL JOINT

The kicker can be formed, in the first pour, as a recess on the form face with the side stop-end penetrated for reinforcement. A beam, bolted to the walers then carries the studs that support the end stop-end. (Figure 4.63)

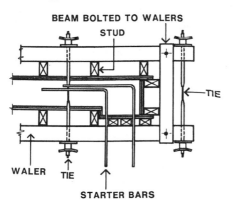

Figure 4.63 - FORMWORK FOR HORIZONTAL KICKER

Even though a horizontal kicker has been provided it is useful to clamp the next form tightly up against the previously cast work. Figure 4.64 shows an effective method of doing this by using the ties from the earlier work.

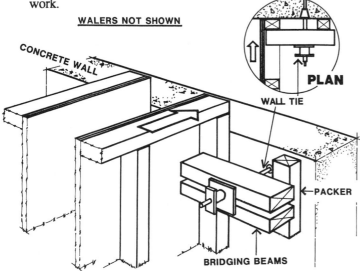

Figure 4.64 - CLAMPING TO EXISTING CONCRETE WALL

Kickers and Kickerless Construction

The purpose of kickers and the some details of their construction were given in Chapter 3. A horizontal kicker for a wall construction joint was shown in Figures 4.62 and 4.63 above.

Kicker construction on narrow footings or the edges of slabs has been shown to be relatively simple. However, where walls are to be constructed in the middle areas of slabs, the achievement of an accurately aligned and level kicker to give a grout tight joint is more complex. Figure 4.65 shows one method.

The kicker forms are made of two longitudinal timbers held together by spacers. The vertical position is fixed by supporting the form on wall tie rods fitted with he-bolt cones. The cones are fixed in place with bolts and washers. The problem with this method is the ease with which the form set can be knocked out of line or tilted by

the concrete placers. Some control of the line is achieved by wires looped from the kicker forms to the reinforcement. Nevertheless, to get an accurate kicker, great care is needed during concrete placing.

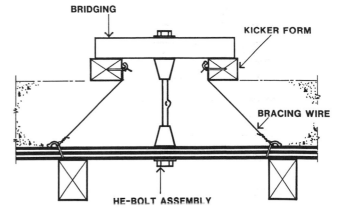

Figure 4.65 - KICKER WITHIN A WIDE SLAB

Because of these difficulties, many walls located within slab areas are built without kickers. For success, two matters have to be addressed: grout tightness and alignment.

With the straight bottom edge of the wall forms seated directly on the trowelled surface of the slabs, there will almost always some small gap between the two. Even the most accurate trowelled surface will have some deviations. Inevitably, there will be some grout loss and boney concrete. Where this is not acceptable the gaps must be filled. One method is the use of a low density plastic foam strip which will squash down to accomodate the variations in the gap. This is included in Figure 4.66 below.

If the foam strips are placed loosely on the slab surface they will be easily displaced while the form is being erected. They should be bonded either to the slab or the bottom plate of the formwork.

The second problem of achieving accurate alignment can be dealt with in a number of ways. At the simplest, battens can be shot-set onto the slab as shown in Figure 4.66. A disadvantage of this method is that no later adjustment of the accuracy of the line can be readily made after the battens are fixed. Also the shot-pin marks usually have to be repaired.

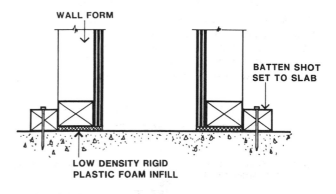

Figure 4.66 - KICKERLESS CONSTRUCTION - 1

Another non-adjustable method is the shot-setting of galvanised sheetmetal angles to the slab as wall form spacers. (Figure 4.67)

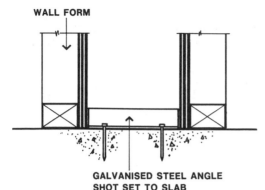

Figure 4.67 - KICKERLESS CONSTRUCTION - 2

Figures 4.68 and 4.69 show adjustable spacer assemblies which are cast into the slab. Apart from the details of the welded support frames, they are similar. Each has a threaded rod fitted with he-bolt cones at its ends. The cones are screwed either way to achieve an accurate line and wall thickness. After stripping the forms, the cones can be removed. To ensure that grout does not fill the hexagonal recesses in the cones, these should be taped or plugged before erecting the forms.

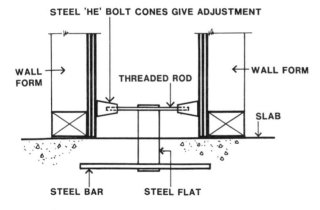

Figure 4.68 - KICKERLESS CONSTRUCTION - 3

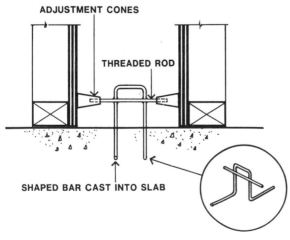

Figure 4.69 - KICKERLESS CONSTRUCTION - 4

Figure 4.70 shows a method of form alignment that is adjustable and can be installed after the slab pour. Circular fibre cement disks are screw fixed through eccentrically placed holes into expanding anchors or plugged holes drilled in the slab. The screws are tightened after rotation of the disks to the correct formwork alignment.

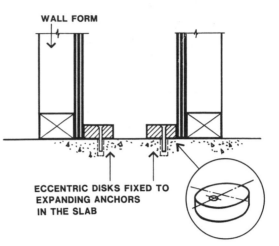

Figure 4.70 - KICKERLESS CONSTRUCTION - 5

Another method for installation after the slab pour, wedges the form against steel pegs that have been inserted in holes drilled in the slab. (Figure 4.71) The pegs are easily withdrawn after formwork stripping and the holes patched. For this method to be satisfactory, the wall ties must be of a type that holds the two forms apart. She-bolts could not be used here without spacers within the wall.

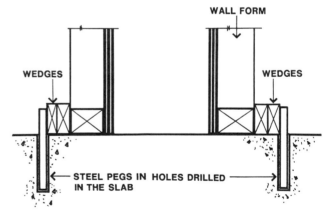

Figure 4.71 - KICKERLESS CONSTRUCTION - 6

As there is no kicker to tighten against, the lowest waler can be placed above the top of the steel pegs. Ideally, there should be sufficient gap to permit removal of the pegs without the need to remove the bottom walers.

Kickerless construction is not confined to the interior of the slab. Formworkers often choose to build walls on the slab edge without kickers. Figure 4.72 shows a spandrel wall formed on the edge of a previously built slab and beam floor.

When the floor slab and its edge beam are poured, pigtail anchors are cast into the beam as fixings for the outer wall form. A grout tight joint is readily achieved on this outer face but the bottom of the inner form, not being clamped to a kicker, will have the moisture and grout loss problems noted previously.

If he-bolt or bar-tie wall ties are used, no extra provision has to be made to control the line of the inner form face.

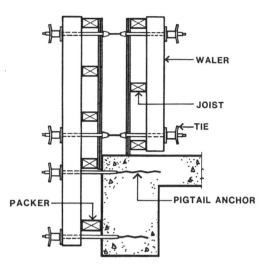

Figure 4.72 - SPANDREL WALL FORMS WITHOUT KICKER

External Corners of Wall Forms

Where two walls meet at a corner the formwork to the outside, at the external corner, requires special attention. Figure 4.73 is a plan view showing the problem in principle. The wall forms on the outside cantilever beyond the last line of ties. Unless provision is made to prevent this cantilever action, the forms will deflect apart and severe grout losses will occur.

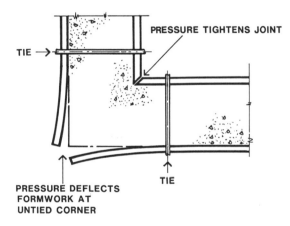

Figure 4.73 - EXTERNAL CORNER OPENING

For minimum grout loss at the corner, there must be a positive tying force to hold the two formwork assemblies together at their junction.

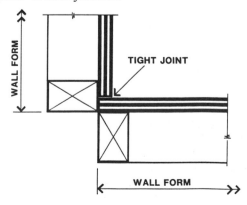

Figure 4.74 - PLAN OF CORNER JUNCTION

Grout tight formwork corners start with a tight junction between the two plywood faces. Figure 4.74 shows the corner interface which is also commonly used in column formwork.

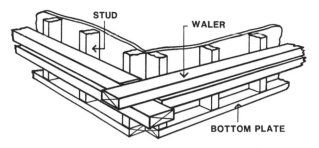

Figure 4.75 - WALER LAPPING AT THE CORNER

A basic detail which is common to many of the methods for positively clamping the corner tight is shown in Figure 4.75. A cutaway view of the corner of horizontal waler formwork shows the half lapping of the double walers at the corner. For vertical waler forms, the horizontal joists can be extended in a similar manner and most of the details given below can also be used.

Eight methods for corner tying, shown in Figures 4.76 to 4.82 inclusive, are based on the extension and lapping of the longitudinal timbers at the corner. The first and simplest of these is given in Figure 4.76.

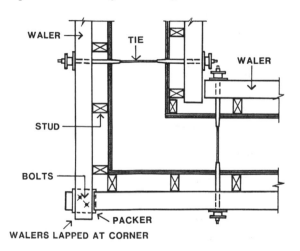

Figure 4.76 - WALL FORM CORNER TYING - 1

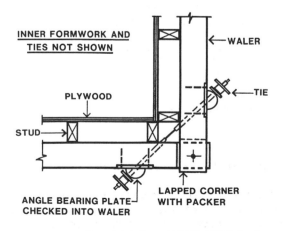

Figure 4.77(a) - WALL FORM CORNER TYING - 2

While the bolting can effectively prevent large corner deflections when the bolts come under load, there will be a strain movement that can result in a small opening action at the plywood corner junction. Grout loss can result. For best results some means of pre-tightening the corner is needed.

One method is the addition of a proprietary angle tie rod placed diagonally across the corner and fitted with angle bearing plates. These angle bearing plates are checked into the walers to prevent them sliding. (Figure 4.77(a)).

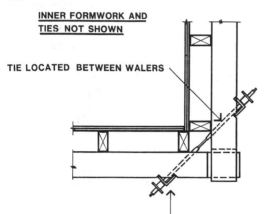

Figure 4.77(b) - WALL FORM CORNER TYING - 3

Figure 4.77(b) shows a similar method with standard wall ties used with fabricated steel angle cleats notched into the walers.

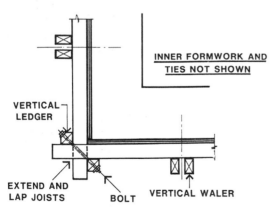

Figure 4.78 - WALL FORM CORNER TYING - 4

Figure 4.78 shows the plan of the corner of vertical waler wall forms. If the diagonal ties at the corner are installed through vertical ledgers, then the bolting of the horizontal joists is not usually needed.

As Figure 4.79 shows, the forces on the inner forms can be balanced against those on the outer forms. Two vertical ledgers are used, one in the the inner corner and one on the outer. These bear against the horizontal walers. She-bolts and threaded tie rods are placed diagonally between them.

Although this can be an effective method it is not in common use. The holes for the ties are difficult to drill and, as the diagram shows, there is considerable congestion at the inner corner.

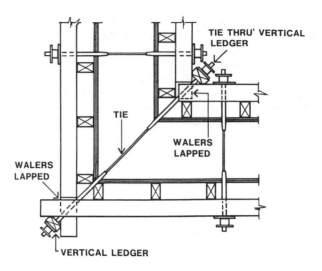

Figure 4.79 - WALL FORM CORNER TYING - 5

The ties that strain the vertical ledger onto the walers can also be placed along the lines of the wall forms. Figure 4.80 shows a popular method using telescopic props. With the pin placed on the underside on the adjusting collar, the prop becomes an effective tension member. The prop cap plate hooks over a vertical ledger at the corner and the baseplate hooks over a stud. While this may appear to be clumsy, it is a very effective and frequently used method.

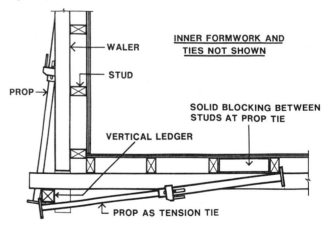

Figure 4.80 - WALL FORM CORNER TYING - 6

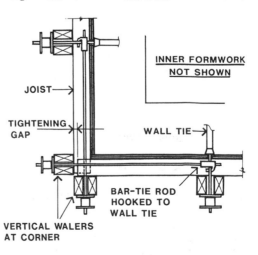

Figure 4.81 - WALL FORM CORNER TYING - 7

The corner tying method shown in Figure 4.81 uses a similar technique to the stop-end of Figure 4.57. Two pairs of vertical walers are linked by bar-ties to the wall ties further along the wall form. The joists only lap a short distance past each other at the corner. They are cut off approximately 20 mm short of the waler's inner face to give clearance for tightening the ties.

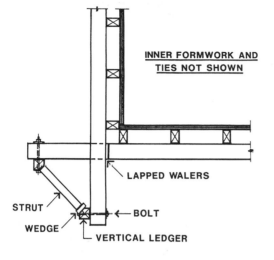

Figure 4.82 - WALL FORM CORNER TYING - 8

The range of hardware needed for the details shown above is not always available. Figure 4.82 shows a method that only involves bolts and timber. The walers are extended with vertical ledgers bolted to their ends. Timber struts are then wedge tightened between the ledgers to deflect the walers and tighten the corner.

All of the above eight cases have shown right-angle corners in walls. Where the junction of the walls have a more acute angle, the cantilever length of the horizontal walers is proportionately greater. This greatly reduces the effectiveness of the cantilever ends in tying the corner. In these cases a tie should be placed across the corner through the wall.

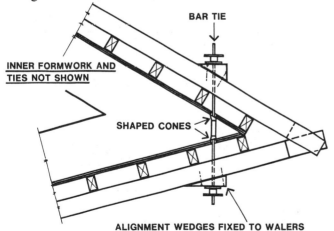

Figure 4.83 - ACUTE ANGLE CORNER TYING

As Figure 4.83 shows, the tie is not at right angles to the wall. Bar-ties are usually best for these as the plastic cone can either accomodate the misalignment or can be cut to line to give a snug fit to the formface. On the outside of the walers, timber wedges are used to give a square bearing to the wall-tie wingnuts.

Internal Corners

The self-tightening action resulting from the concrete pressure on the forms at the internal corner was noted in Figure 4.73. If grout tightness is essential, the sectional plan detail shown in Figure 4.84 is suitable. The fluid concrete pressure causes a closing action at the corner, assisting grout tightness and alignment. For a grout tight joint, a closed cell polyurethane tape can be used to fill all the minor irregularities. The foam tape must be fully compressed. The gap between the faces of the two studs is needed to ensure that they do not come into contact and prevent this compression.

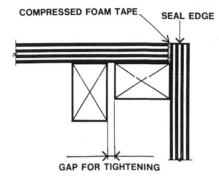

Figure 4.84 - INTERNAL CORNER TIGHTENING

Where only an average standard of grout tightness at the internal corner is needed, the detail given in Figure 4.85 can be used. Because the plywood sheets do not meet at the corner, it has the added advantage of ease of stripping. A sheetmetal angle, of at least 1.2 mm thickness, is fixed to one form (usually with 15 mm clouts) and lapped onto the other. The concrete pressure holds it tight to both plywood faces and, with care, an acceptable degree of grout tightness results.

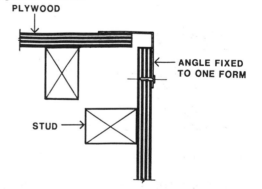

Figure 4.85 - METAL ANGLE AT INTERNAL CORNER

'Tee' Junctions Between Walls

Forms become more complex and costs rise when a monolithic junction is called for at a 'Tee' junction between two walls. Figure 4.86 shows one of the methods of constructing the forms at the junction.

Care has to be taken with the alignment of the wall formwork abutting the side of the other wall. The extension of the ends of the plywood is usually quite flexible. To control the bottom, a kicker, or equivalent, is needed at the base of the wall forms. The line of the top, and at mid-height, can be controlled by a plywood gusset or brace across the top line of walers.

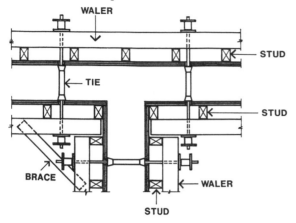

Figure 4.86 - PLAN OF FORMS AT A 'Tee' JOINT

This formwork is complex and involves cutting the wall forms to fit either side of the wall junction. As discussed in Chapter 1, for economy, this should be avoided where possible.

If a construction joint is permitted at the junction, the two stage method shown in Figures 4.87(a) and (b) can be used to save cutting straight wall forms. The 'Tee' shaped wall is poured in two parts.

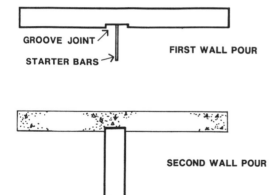

Figure 4.87(a) - CONSTRUCTION JOINT FOR 'Tee' JOINT

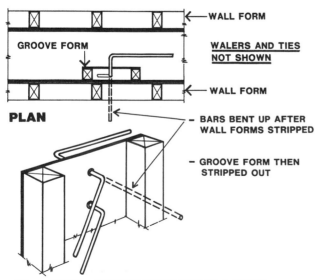

Figure 4.87(b) - GROOVE FORM FOR 'Tee' JOINT

This technique is an adaption of that used for the connection of floor slabs to slip-formed walls. It enables starter bars to be installed for joining the 'Tee' wall without the formface being defaced with holes.

A groove form is inserted in the wall on the formface. This contains reinforcement bars which, after the stripping of the wall form, are bent out to be starter bars to the joining wall. The groove form cannot usually be recovered for re-use. Its removal normally results in its destruction.

At the second pour, the detail given earlier in Figure 4.64 can be used to clamp the forms for stage two, tightly to the first wall.

Surface Features on Wall Faces.

It is common for walls to have surface features, such as grooves or recesses, or built in fittings such as welding plates.

Grooves and recesses are usually specified for aesthetic reasons, and as a result must be formed to have clean sharp lines with faces and corners free of defects. To ensure that damage does not occur during striping, the form must be appropriately shaped and all surfaces must be smooth and sealed.

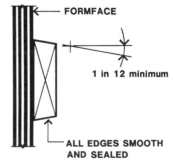

A Decorative Pattern of Surfaces Recesses

For ease of formwork withdrawal, the sides of all groove and recess forms should have a minimum slope of 1 in 12. (Figure 4.88)

Figure 4.88 - TAPER TO GROOVE FORMS

Grooves are often in a repetitive pattern within a single pour. If there are a great number, and closely spaced, they can make stripping very difficult. The application of the force needed for stripping can often result in fracture of the outstanding concrete. Two things can be done. Firstly, as shown in Figure 4.89, the concrete 'land' between the grooves should be at least one and one half times as wide as the groove depth to minimise fractures. With a wider concrete section, the concrete oustand will have greater strength.

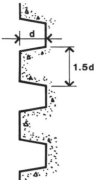

Figure 4.89 - PROPORTIONS FOR MULTIPLE GROOVES

Secondly, a system that minimises the groove form extraction force should be adopted. Figure 4.90 illustrates one method. The grooves are formed by tapered section metal or plastic channels which are firmly seated over timber locating strips on the formface. During formwork stripping the timber locating strips, which are fixed to the formface, readily pull out of the channels. The channels are stripped separately by individually flexing their sides inwards by a small amount.

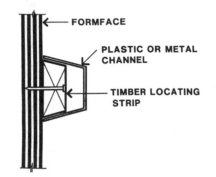

Figure 4.90 - TWO PART GROOVE FORMS

Grooves are often used to partially conceal construction joints, both vertical and horizontal. For horizontal grooves, the construction joint can be located near the top of the groove so that it tends to lie in shadow and is less conspicuous. (Figure 4.91)

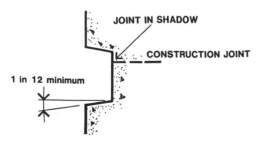

Figure 4.91 - CONSTRUCTION JOINT IN A GROOVE

When forms with horizontal grooves are used for a reasonable number of repetitious pours, one above the other, then errors in alignment can occur. A simple example of walls with horizontal grooves at all construction joints illustrates the problem.

The wall form will be provided with a groove form at its base, for seating in the groove cast into the top of the previous pour. Also there will be a groove form at its top to form the groove in the current pour. However, even the most accurately constructed form can have minor fabrication errors. These are illustrated in Figure 4.92. This shows the elevation of the wall form with top groove form slightly out of parallel with the lower groove form.

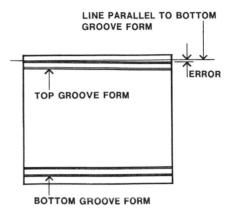

Figure 4.92 - ELEVATION OF WALL FORM

Figure 4.93 shows how this small error will enlarge to a significant progressive error.

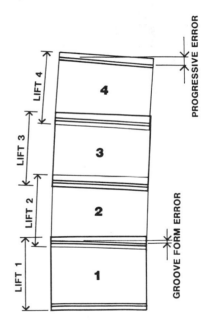

Figure 4.93 - PROGRESSIVE ERROR WITH FOUR USES

To avoid this progressive error problem, a full double tapered groove form should not be used; it 'locks' the second and following lifts of the form into an unchangeable position. There has to be a means of adjusting the level of the top groove at each lift of the form. This is achieved by changing the groove forms.

At the top of the wall form the groove form only needs one tapered edge. Its top can be square and provide a screeding batten. (Figure 4.94 A) A similar groove form, but reversed in position, is shown used at the bottom of the wall form. The width of its flat face is less than the face width of the groove being formed. This enables adjustments to the level of the form. (Figure 4.94 B)

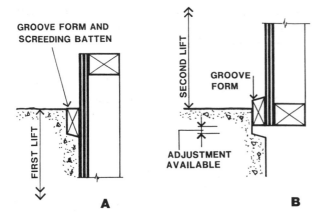

Figure 4.94 - GROOVE FORMS FOR ADJUSTMENTS

Square sided grooves are sometimes called for. A single piece groove form would be nearly impossible to strip. The stripping difficulties can be reduced by using a two piece form as shown in Figure 4.95. Both parts are screw fixed onto the formface (Figure 4.95 A). To ease stripping the screws to one part are removed, and the wall form removed with only one part of the groove form attached (Figure 4.95 B).

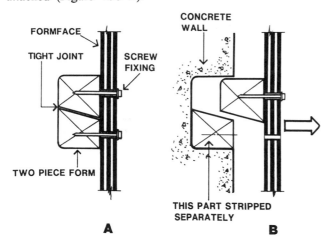

Figure 4.95 - TWO PART SQUARE GROOVE FORM

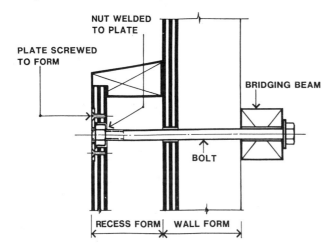

Figure 4.96 - SECTION OF DEEP RECESS FORMWORK

Forms for large and deep recesses in walls can pose stripping problems. The peeling action is difficult due to the torsional stiffness of a deep form, and problems in easing one corner out without defacing the concrete. For direct formwork removal away from the concrete face, special devices must be provided. Figures 4.96 and 4.97 show one scheme for achieving this.

The recess form is bolted to the wall form. The bolt connection is through a nut welded to a plate, which in turn is sunk flush into the formface and screw fixed. To strip the wall form the bolts are removed so that the recess form stays against the concrete.

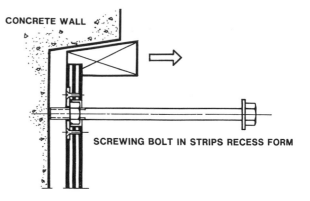

Figure 4.97 - STRIPPING THE DEEP RECESS FORM

To strip the recess form, after separately stripping the wall form, the bolts are wound back into the nut and plate assemblies to force the recess form out. (Figure 4.97)

It is often necessary to build in fittings and fixings at the wall face. Figure 4.98 shows a plate and anchors cast into a wall for the field welding of a structural member to the wall.

The fitting is usually held onto the formface by bolts through the plywood. Threaded holes or nuts welded to the fittings are provided for these bolts.

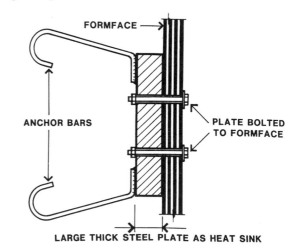

Figure 4.98 - FITTINGS ON THE CONCRETE SURFACE

When the plate is to be used for field welding a very thick plate is always used. This is to provide a 'heat sink' to reduce the average temperature of the plate during welding. A thinner plate would have concentrations of heat at the weld, and cause spalling of the concrete nearby.

Bolts, for building into the wall, are also fixed to the formface. Figure 4.99 shows a typical case of a bolt held in line by a nut on each side of the formface. The inner nut and washer remains in place after the form is stripped.

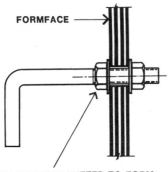

Figure 4.99 - CAST IN BOLT

With a number of bolts protruding from the concrete face the stripping of the formwork can often be difficult. This problem can be avoided by using threaded ferrules. As shown in Figure 4.100 'A', a ferrule is fitted with an anchor and bolted to the formface.

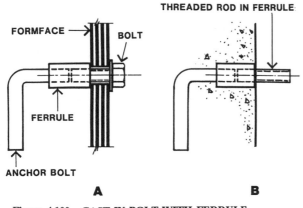

Figure 4.100 - CAST IN BOLT WITH FERRULE

The bolts are removed after the pour, and then there are no projections to impede stripping. After formwork stripping, a length of threaded rod can be screwed into the ferrule to provide the equivalent of a bolt projecting from the concrete face. (Figure 101 'B')

Penetrations and Openings Through Walls

Penetrations through walls range in size from small openings for pipes and conduits to large door openings. In all cases the construction objectives of penetration forms are:
- accuracy of size and position,
- adequate fixing to resist flotation and sideways movement due to concrete pressure,
- ease of stripping,
- recovery of penetration formwork where possible.

Penetration forms, being surrounded by concrete, are more difficult to strip than most forms due to the shrinkage of this concrete. Further, where the forms are made of timber, water penetration may cause swelling of the timber which would increase the stripping problems. For rectangular openings, the construction can be arranged to make dismantling of the form easier. Figure 4.101 shows a typical case.

Internal blocking holds the form to shape and this blocking is also fixed to one formface of the wall formwork to maintain its position. The outer form faces of the penetration are lapped over each other at successive corners for ease of stripping.

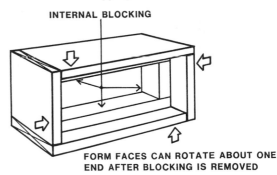

Figure 4.101 - RECTANGULAR PENETRATION FORM

Stripping starts with the formwork for the wall face opposite to the one to which the blocking is fixed. The blocking is then accessible and is stripped out. The formwork to the other wall face can then be stripped. As Figure 4.102 shows, the outer faces of the penetration forms can be rotated for removal.

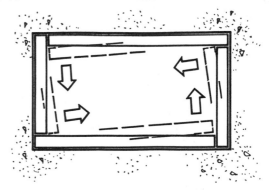

Figure 4.102 - PENETRATION FORM REMOVAL

The larger forms needed for door openings can be built in the same way, but due to their size a different technique for compensation for tightening due to concrete shrinkage can be used. This is shown in Figure 4.103.

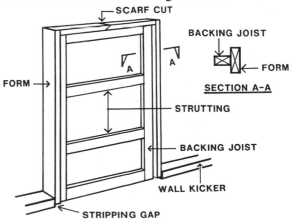

Figure 4.103 - PENETRATION FORM FOR A DOOR

Similar to the other case, the interior is braced with backing joists and strutting, the struts being fixed to one formface to control the penetration form position. The top formface is made with a scarf cut in it. During stripping this internal framing is readily removed. The shrinkage induced tightening is relieved by the scarf cut. The provision of a small stripping gap (about 3 mm) at the bottom of the side forms makes their removal easy.

Where neatness and absolute accuracy of position are not crucial, a simpler method can be used for small rectangular penetrations. Figure 4.104 shows a section through the wall penetration.

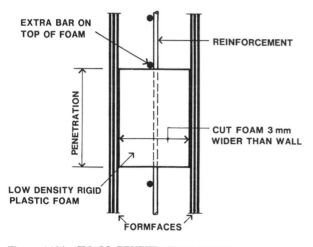

Figure 4.104 - SMALL PENETRATION FORM

These penetration forms can be made from blocks of low density rigid plastic foam. They are held reasonably closely in position by tying to the wall reinforcement. To stop flotation an extra reinforcing bar can be place on top and tied firmly to the wall reinforcement. Friction with the formfaces can be achieved by making the block of foam approximately 3 mm wider than the wall. The clamping action of the wall ties then compresses the foam to create this frictional resistance. Usually, no material recovery is possible when this form is stripped, as the foam has to be broken up for removal.

For pipes, drains and airconditioning ducts, circular penetrations are usually needed. Small holes can be formed with a length of plastic tube suspended on a rod as shown in Figure 4.105. The tube should be cut slightly longer (say 1 mm) than the wall thickness so that tightening the wall ties helps to hold the tube in position. Nevertheless, some movement may occur.

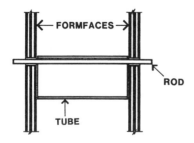

Figure 4.105 - SMALL PIPE PENETRATION

To more accurately control its position, three rods can be used, or a locating block can be fixed on each formface. A development of this is shown in Figure 4.106. Here the locating block is made of rubber and bolted to the formface through a circular washer which is only slightly smaller than the tube. Tightening the bolt

compresses the rubber which expands against the inside of the tube. If the outside of the tube is suitably coated with mould cream, or other heavy bodied release agents, it can usually be recovered.

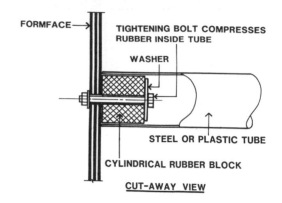

Figure 4.106 - ACCURATE PIPE PENETRATION

Larger round penetrations can be formed with tubular void forms. Figure 4.107 shows the use of cardboard tubular void forms. These are made from multiple layers of cardboard laminated to approximately 10 - 12 mm thickness, and waxed on external faces to resist moisture. The ends, however, are susceptible to water penetration which will weaken the cardboard. The ends should be treated or taped to prevent this. The circular shape and position is maintained with discs of plywood fixed to the formfaces. As with most cardboard forms it is destroyed during removal.

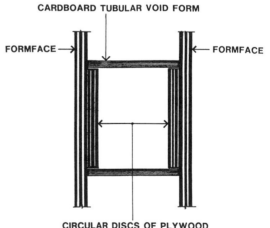

Figure 4.107 - CARDBOARD TUBE PENETRATION FORM

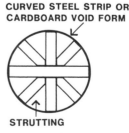

Figure 4.108 - LARGE CIRCULAR PENETRATION

Circular penetrations can often be required in quite large sizes. For these the shape can be maintained by strutting. (Figure 4.108) The circular edge can often be formed by cardboard void forms or, for larger sizes, a curved metal strip.

Controlling the Top of the Wall Pour

Accuracy of the line of the top of the concrete pour is usually as important as the accuracy of the line, level and plumb of the wall formwork. Further, the method used to control the top of the wall line should be able to be adjusted for each use of the formwork. This will enable adjustment for progressive errors and it will not impose the requirement that the forms should always be erected at a precise predetermined level.

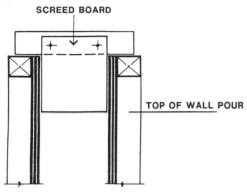

Figure 4.109 - CONTROLLING THE TOP OF THE POUR - 1

Figure 4.109 shows one of the simplest methods. With the forms erected with their top line parallel to the wall top, usually level, a simple screed board can be made. However, this gives rough top edges to the wall. If neat square corners are needed the detail shown in Figure 4.110 can be used.

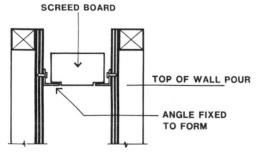

Figure 4.110 - CONTROLLING THE TOP OF THE POUR - 2

Angles are fixed to line on the formface and the screed board is shaped with small recesses that match the angle thickness. For a smooth, though recessed, top surface, a trowel can be used along the angles.

If the top of the pour is a construction joint and a smooth face with a neat level junction line is needed, then, the detail shown in Figure 4.111 is more suitable. This diagram shows the formwork and a section through the resulting construction joint.

Square battens, about 25 x 25 mm, are fixed to the formfaces and a screed board made to set the top of the pour at about half the batten height. Adequate vibration of the concrete ensures a sharp neat concrete line along the underside of the batten. Usually, the screeding between the projecting vertical reinforcement does not have to be particularly accurate.

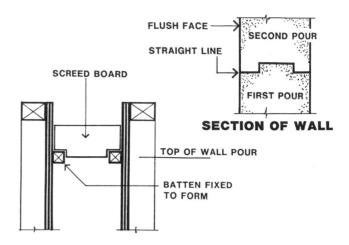

Figure 4.111 - CONTROLLING THE TOP OF THE POUR - 3

When a keyed shape is specified, for a construction joint at the top of a wall, a key form which is fixed in position will obstruct effective concrete placement and vibration. To avoid this problem a separate unfixed form for this key can be used (Figure 4.112).

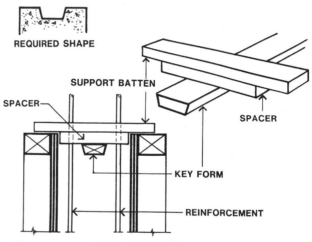

Figure 4.112 - KEYED CONSTRUCTION JOINT

The concrete is placed up to approximately the underside of the level of the key. Lengths (say 2 metres) of the key form and its supporting battens are then placed and nailed in position before the final part of the concrete is poured. The spacers fix the accuracy of its line. Nailing is needed to stop flotation of the key form during vibration of this last layer of concrete. The screed board for this detail is similar to that shown in Figure 4.109.

However, this key form can also be used in conjunction with the details given in Figures 4.110 and 4.111 and screed boards similar to the ones shown with those details would be used.

Access Platforms

Efficiency in the work of placing and vibrating concrete in the wall forms calls for a safe and adequate access platform for workers and equipment. This can be an independent scaffold alongside the forms. To limit any relative movement the scaffold should be tied to the wall formwork. Alternatively, these access platforms can be built as part of the formwork framing.

Access platforms for fixing as cantilever frames onto the wall formwork framing can be proprietary units, for which a number of types are available, or they can be purpose built. In all cases it is important that the framing of the cantilever platform is adequately fixed to the major framing members of the formwork. For a horizontal waler form one suitable method is shown in Figure 4.113.

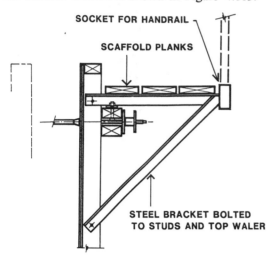

Figure 4.113 - HORIZONTAL WALER FORM & PLATFORM

The top member, which is subject to tension, is shown bolted to both vertical studs and the top waler. The compression strut is bolted to the studs. To ensure a sound structure, the waler should be cleated to the studs. The platform framing must be fitted with attachments for a continuous guardrail at its outer edge.

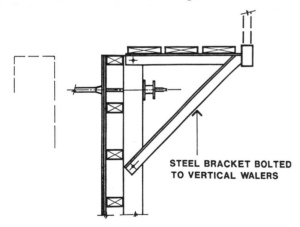

Figure 4.114 - VERTICAL WALER FORM & PLATFORM

A platform for a vertical waler form is shown in Figure 4.114. Both platform members are shown fixed to the vertical walers. The vertical walers should be positively fixed to the joists. These fixings can be direct bolting of joists to walers or bolting through angle cleats.

Provision for Hoisting Wall Forms

With the proliferation of cranes on building sites, only small wall forms are man-handled into position. For all other cases the special requirements of designing the formwork for hoisting have to be taken into account.

Two matters have to be considered: firstly, the fixings within the form structure, to give it adequate strength and stiffness for hoisting, and, secondly, the provision of properly designed hoisting points.

In both situations the manner of hoisting must be considered. It is best, for hoisting purposes, to store wall forms on their bottom edges in a special rack, as shown in Figure 4.115. However, they are often laid on their flat and have to be hoisted from that position. All the fixings must cater for this type of loading.

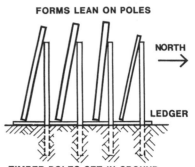

Figure 4.115 - STORAGE RACK FOR WALL FORMS

Standing forms on their edge has two further advantages. Firstly, they will not be damaged by foot traffic, and secondly, they can be faced away from the sun so that this source of deterioration is avoided.

The need for sideways form stiffness was illustrated in Figure 4.02 and, later, reference was made to the importance of lapping the plywood sheets (Figure 4.39) and the selection of adequate plywood fixings. In general, nails are not adequate to resist the repetitive racking action that occurs with crane hoisting and the wind loads that can simultaneously occur. Screw fixings as shown in Figures 4.48 and 4.49 should be used. In most cases the spacing of the fixings should not exceed 200 mm along the edges of the form and 300 mm internally. For very large formwork units closer screw spacings will be needed.

Fixings between framing members; studs, joists, plates and walers, must also be adequate. Figure 4.116 shows plywood gussets for the corner of the forms. Figure 4.117, on the next page, shows a similar gusset for use at internal studs.

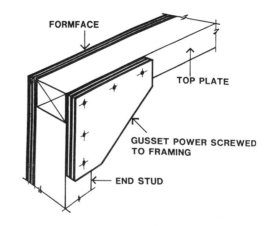

Figure 4.116 - GUSSETS FOR CORNERS OF FORMWORK

For forms which are to be crane hoisted, the walers must also be effectively fixed to the sub-framing, studs or joists. At the least coach screws or bolts are needed. If the hoisting is applied to the walers then bolted angle cleats, walers to studs, will be needed as a minimum. This is shown in later illustrations.

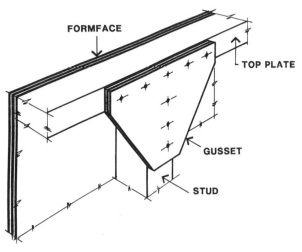

Figure 4.117 - GUSSETS FOR INTERNAL FRAMING

For long wall forms the hoisting setup should be arranged to minimise the bending effect on the form. Firstly, the hoisting points should be spaced at about 60% of the form length. Secondly, long lifting slings should be used to minimise the compression induced in the top of the form by the inclined sling forces. (Figure 4.118) If shorter slings are used this compression force increases. Buckling of the top of the form may occur.

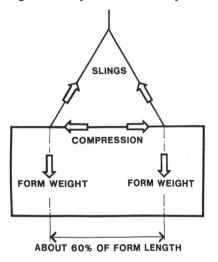

Figure 4.118 - HOISTING ARRANGEMENTS

Two examples of hoisting connections for this case are given. Figure 4.119 shows the lifting eyes bolted to both studs and plywood of a horizontal waler form. Countersunk headed bolts are used to give a flush formface. The walers are not part of the support structure, but by being coach screwed or bolted to the studs, they provide longitudinal stiffness to the overall form assembly. The size and type of this fixing of the walers to the studs must match the strength and stiffness of the walers.

The stiffness of the walers becomes important during lifting the forms from the horizontal to the vertical. The walers resist the tendency of the form to bend as the hoisting starts. Also, they help resist flexure under the effects of wind loads while being hoisted.

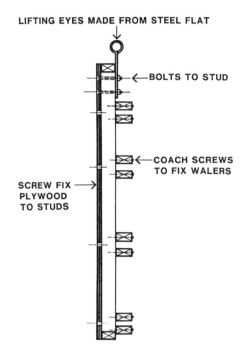

Figure 4.119 - WALL FORM HOISTING FITTINGS - 1

A similar hoisting fitting can be used for vertical waler forms.

Figure 4.120 shows the lifting points connected to the horizontal walers. The continuously threaded rod is shown passing through three pairs of walers with steel angle connections at each waler pair. The angles are also bolted to the plywood and studs with countersunk headed bolts. Through these angle connections, the walers are effectively connected to the studs at the lifting points. At studs, between and on each side of the lifting points, the walers should be coach screwed or bolted to the studs.

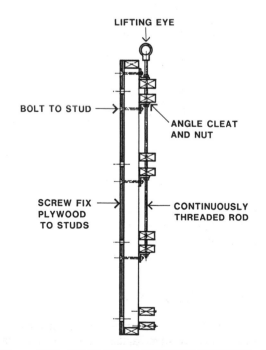

Figure 4.120 - WALL FORM HOISTING FITTINGS - 2

Shorter wall forms, where, for example, the length is no more than one and one half times the height, can usually be hoisted at their ends, or provided with a stiffening beam for slinging at any position. Figure 4.121 shows this latter case. A steel channel is shown bolted to all studs for the full length of the form and, like the other cases, the walers are coach screwed to the studs to give lateral stiffness to the form.

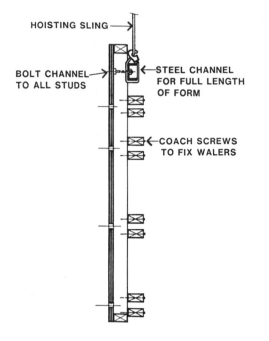

Figure 4.121 - WALL FORM HOISTING FITTINGS - 3

A suitable detail for hoisting points at the end of a short wall form is shown in Figure 4.122. As the lifting plates are bolted only to the end studs and have no direct connection to the plywood face, extra screw fixings through the plywood to this end stud are needed. As shown, the framing should be gusseted at the corner.

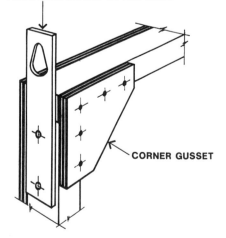

Figure 4.122 - WALL FORM HOISTING FITTINGS - 4

Where a total wall form assembly consists of a sequence of short wall forms, the use of these end lifting fittings must be catered for with the provision of a gap between forms for the insertion of the lifting hooks or shackles. Figure 4.123 shows a plan view of the junction between two forms.

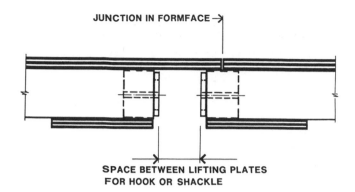

Figure 4.123 - PLAN AT JUNCTION OF FORMS

Pilaster Piers

Pilasters are piers attached to a wall and can be regarded as columns incorporated in the wall. As Figure 4.124 shows they can project from one or both sides of the wall.

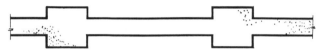

PILASTERS PROJECTING FROM BOTH WALL FACES

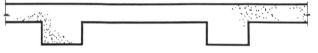

PILASTERS ON ONLY ONE WALL FACE

Figure 4.124 - PILASTER TYPES

Their construction monolithically with the wall causes formwork complexities. Wall forms have to be tailored to fit between the faces of the outstanding piers and the formwork to the pilasters has to fitted between the ends of these wall forms. The plan view of Figure 4.125 shows this situation for shallow pilaster piers.

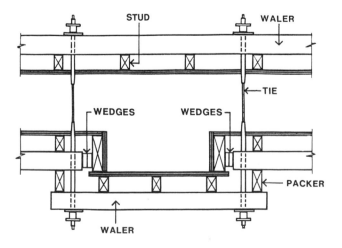

Figure 4.125 - FORMWORK FOR SHALLOW PILASTERS

For deep pilasters, Figure 4.126 shows the formwork to be even more complex.

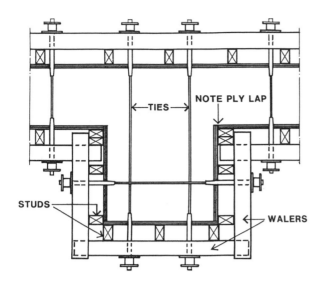

Figure 4.126 - FORMWORK FOR DEEP PILASTERS

These difficulties can be generally avoided if the pilasters can be poured separately to the wall. Figures 4.127 and 4.128 illustrate the sequence.

PLAN OF WALL WITH PILASTERS

POUR WALLS FIRST

THEN POUR PILASTERS

Figure 4.127 - CONSTRUCTION SEQUENCE - PLANS

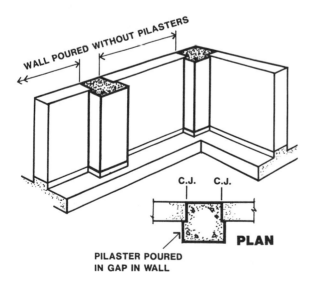

Figure 4.128 - CONSTRUCTION SEQUENCE - PERSPECTIVE

There are significant cost savings in this procedure. Special short wall forms do not have to be made or cut

from longer forms. These long wall forms can be fully recovered from the wall pour for future repeated uses. In this way the unit costs of formwork are markedly reduced.

The gaps in the wall for the pilasters are simply a double stop-end with wall ties located alongside to suit their later use with the pilaster forms. The width of the gap for the pilaster should be made approximately 20 mm narrower than the pilaster. This gives some tolerance in the location of the gap and makes the construction joint less obvious when the pilasters have been completed.

Because it is likely that the depth of fluid concrete on one side of the stop-ends will not always match that on the other side, the stop-ends can be subject to the full concrete pressure from either side. The fixings of the stop-ends should follow the previously given requirements. Figure 4.129 shows the strutting, between the stop-ends, bolted to the formfaces. In most cases these stop-ends will have to accomodate horizontal wall reinforcement passing through.

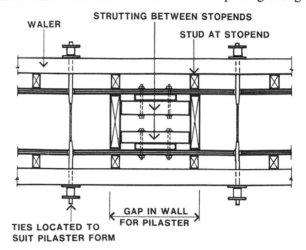

Figure 4.129 - DOUBLE STOP-ENDS AND STRUTTING

After the walls are poured and stripped the pilasters can be constructed. Figure 4.130 shows the formwork for symmetrical pilasters. These formwork units are clamped onto the walls with the nearby ties that were previously used for the wall forms.

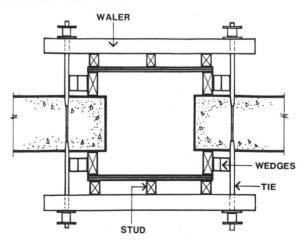

Figure 4.130 - SYMMETRICAL PILASTERS

As noted above, for symmetrical pilasters, the construction joints between walls and the pilasters can be concealed just within the edge of the pilasters. Where the wall and pilaster have a flush face the construction joint is always evident on this face. The formwork for flush faced pilasters is shown in Figure 4.131.

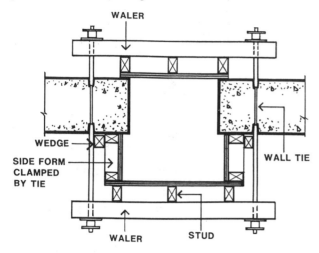

Figure 4.131 - ASYMMETRICAL (FLUSH) PILASTERS

If this is unacceptable then the technique shown in Figure 4.132 can be used.

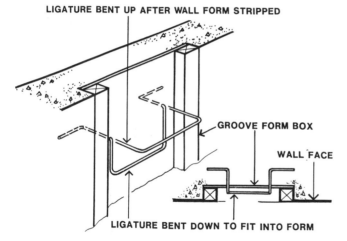

Figure 4.132 - PILASTER GROOVE FORM

The disposable groove form is removed and the ligatures bent out. This is similar to the detail given in Figure 4.87. To construct the pilaster, a form like that shown in Figure 4.131, above, is clamped onto the wall with the wall ties.

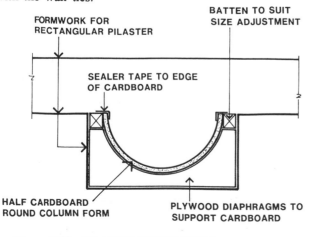

Figure 4.133 - PILASTER INSERT FORM

Non-rectangular pilaster shapes can be formed by supporting the special shaped form within the conventional pilaster form. Figure 4.133 shows the insert form for a half-round pilaster. The details of the formwork to the rectangular pilaster and the walls are not shown.

Variations in Wall Thickness

All the formwork shown so far, has been for walls of uniform thickness. Non-uniform walls with variations such as changes in thickness, haunches and tapers are often needed. Two methods of forming changes in thickness will be shown. Figure 4.134 shows a vertical section of formwork for a wall with a step constructed into one side.

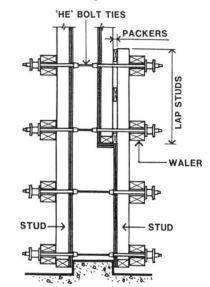

Figure 4.134 - STEPPED WALL FACE - 1

This is a complex solution as the stepped face is really two fabricated formwork units. He-bolts are used to clamp the two units together at their lap. A simpler method, using existing forms without damage, is shown in Figure 4.135.

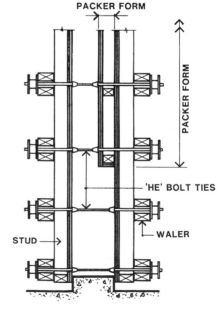

Figure 4.135 - STEPPED WALL FACE - 2

Here the formwork is constructed for a uniform wall of the greater thickness, complete with stop-ends, and a simple packer form equal in thickness to the face step is placed inside, between the stop-ends, and clamped to the main form with the He-bolts.

Figure 4.136 - SECTION OF HAUNCH TO WALL

Haunches on wall faces, like that shown in Figure 4.136, are sometimes needed for the support of slabs, precast floor units or steelwork. If it is necessary to form these monolithically with the wall and somewhere in the middle height, the forms become quite complex and expensive. Figure 4.137 shows how the forms to the haunch side of the wall are made from separate pieces.

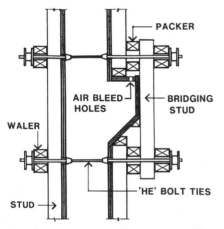

Figure 4.137 - FORMWORK TO HAUNCH

Once again, the clamping action of the He-bolts is used to hold the two parts, of wall forms on this haunch side, to the other plane formwork face. The formwork to the haunch itself, is fixed to them. To ensure that air locks do not prevent the concrete completely filling the haunch, air bleed holes are needed in the top of the haunch form. Fifteen millimetre holes at 400 mm centres are usually adequate.

If it is possible to make a construction joint at the top of the haunch, the formwork is considerably simplified. Figure 4.138 shows one method of forming this situation.

The plane wall form on the haunch side is either shorter than the plane side or set at a lower level. Lapped extension studs are bolted to the wall form studs. A top plate is provided for the support of the upper edge of a folded sheet metal haunch form. The thickness of the sheetmetal must be selected to cater for the concrete pressures. It will rarely be less than 1.6mm thick, and for some larger haunches may as thick as 3 mm. The sheetmetal form is screw fixed along both edges. This type of formwork system can use existing plane wall forms with the minimum of alteration.

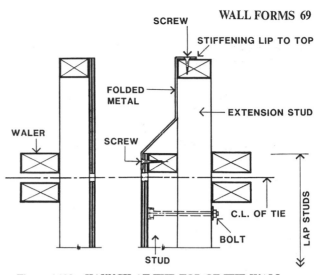

Figure 4.138 - HAUNCH AT THE TOP OF THE WALL

If the haunch can be poured separately, the insert groove form technique with bent out reinforcement can be used. This has already been shown in Figure 4.87 and 4.132. After the wall has been stripped, the bars are bent out and the groove form removed. The haunch form can then be clamped on as shown in Figure 4.139. Although all wall tie types can be adapted to this detail, bar ties are the most suitable, because of the ease of use of two wingnuts on the same tie.

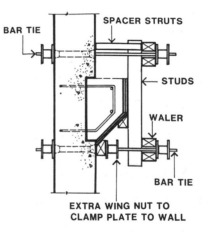

Figure 4.139 - SEPARATE HAUNCH FORMWORK

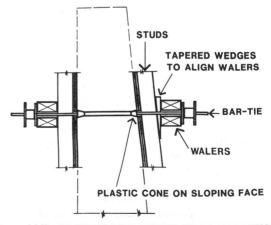

Figure 4.140 - SECTION OF TAPERED WALL FORMWORK

Bar ties also have advantages if used in tapered walls. Figure 4.140 shows a part section of horizontal waler formwork for a tapered wall. The junction of the wall ties with the tapering face are not right angles. However, except for severe tapers the plastic cones are able to distort to cater for this angle.

Access Panels

In the construction of most walls the concrete is placed from the top and the thickness of the wall and the disposition of its reinforcement should be arranged to permit the efficient pouring and compaction of the concrete. However, in some cases variations in the concrete shape, or complex and closely placed reinforcement, make top placement and compaction of all the concrete too difficult. In these cases access panels for placing concrete directly into the lower part of the wall can be used.

Usually these 'windows' in the formwork are located at approximately half the form height, and between lines of walers.

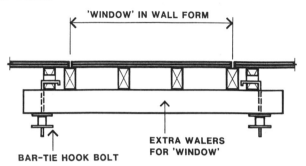

Figure 4.141 - SECTIONAL PLAN OF ACCESS PANEL

Figure 4.141 shows a plan section through an access panel in a horizontal waler wall form. The studs to the access panel, frame onto short walers. These in turn are hook bolted onto holes in the studs. Bar ties are ideal for these hook bolts.

The concrete is placed up to the bottom of the panel opening and then the panel is installed so that the pour can proceed. The junction of the panel with the plywood of the formwork is constructed with a lap to limit grout and moisture losses. Nevertheless, some losses are likely to occur. To enable these losses to be washed away before they harden, adequate hose access should be provided. Because of this potential for losses, 'windows' should not be used where the concrete surface is required to have good visual quality.

PROPRIETARY WALL FORM SYSTEMS AND COMPONENTS

A large proportion of the walls to be formed and poured have dimensions; thickness, height and length, that lend themselves to the use of prefabricated standard modular wall form systems. A number of types of such systems are commercially available.

The oldest of these is the all steel panel system. Although this has been superseded in many cases by lighter panel systems, it is still being used in civil engineering works. For this reason and because many of its principles and details of fabrication and assembly are still used in the more recent developments, it will be described in some detail.

This information is only a general description of this type of product, as they vary in detail between different brands. If using one of these systems, reference should be made to the manufacturer's technical literature for the particular brand.

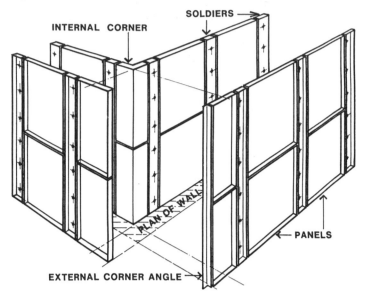

Figure 4.142 - PROPRIETARY WALL FORMWORK

Figure 4.142 shows an 'exploded' view of the part of a formwork assembly for a corner of a wall. The straight portions of the wall consist of a repetitive series of full height steel soldiers, which have holes for the wall ties, with panels between. The structural action of the steel panels is to span horizontally between the soldiers.

These panels are usually available in a range of standard widths and heights. Heights are normally 600, 900 and 1200 mm and widths 900, 600, 450, 300, 200 and 150 mm. At the corner, the inside is formed with a standard internal corner panel, and the outside by straight wall panels joined together by an external corner angle.

The width of the straight corner panel is selected to equal the wall thickness plus the size of the face of the internal corner angle. (Figure 4.143)

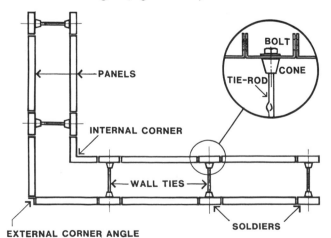

Figure 4.143 - PLAN AT CORNER OF FORMWORK

The ties are of the he-bolt type with a disposable threaded tie rod, recoverable steel cones and bolts to the soldiers. Panels, soldiers and corner angles all have slotted holes along all edges for their connection by screw clamps to adjacent components. Figure 4.144 shows one of these clamps in position. Panels are clamped at both sides of all corners and a number of times between corners, depending on the panel size and the wall height. This important matter is usually covered in the manufacturer's directions.

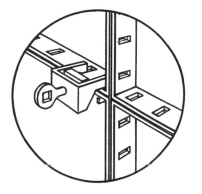

Figure 4.144 - CLAMPING PANEL TO PANEL

Minor variations in the squareness of the panel edges, from either manufacture or field use, can lead to variations in the plan straightness of the form. To ensure a straight line, horizontal steel tubes, of the size typically used in scaffolding, are clamped to the panels as shown in Figure 4.145.

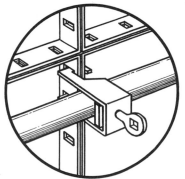

Figure 4.145 - CLAMPING TUBES TO PANELS

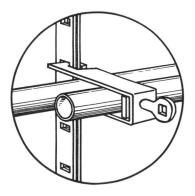

Figure 4.146 - LAPPING ALIGNMENT TUBES

Where the wall form assembly is longer than the tubes, two tubes are butted end to end and a lap tube clamped on top. (Figure 4.146) In general, the function of these tubes is confined to keeping the wall straight, but, there are

some situations where they do carry some of the load from the fluid concrete pressures and other loads.

As noted above, the single panel spans between the soldiers on each of its edges. The load, from the concrete pressure, is balanced through the ties with that from the other side. However, when two panels are used side-by-side between soldiers, usually to suit dimensional requirements, the panels cannot span horizontally. In that case, the tubes are required to carry the load to the soldiers. (Figure 4.147)

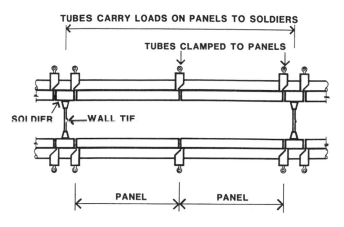

Figure 4.147 - TUBES SUPPORTING PANEL LOADS

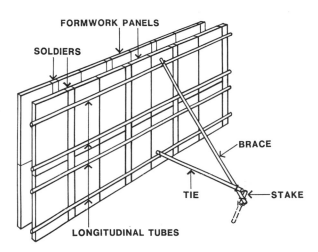

Figure 4.148 - CONNECTION OF BRACES TO TUBES

The tubes are also used to attach wall form braces. In Figure 4.148 they are shown clamped to the top and bottom tubes and to a short tube driven into the ground as a stake.

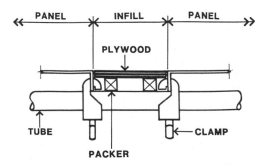

Figure 4.149 - INFILL PANELS

If the standard panels cannot be combined to give the required width dimension, a special plywood infill panel can be made. As the sectional plan of Figure 4.149 shows, vertical packer timbers are used to get face alignment and the horizontal tubes are used to carry the loads to the soldiers.

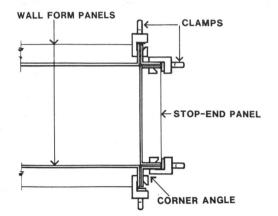

Figure 4.150 - STANDARD PANEL AS STOP-END

Figure 4.150 shows the simple modular stop-end that can be used with these systems. The stop-end panel, equal in width to the wall thickness, is fixed to the edges of the side panels with clamps and external corner angles.

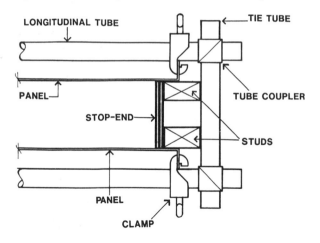

Figure 4.151 - STOP-END AS WALL LENGTH ADJUSTMENT

The stop-end can also be constructed to give a minor adjustment to wall length. As Figure 4.151 shows, for this minor wall length adjustment, the stop-end is carried off the ends of the horizontal tubes. A short tie tube is clamped across the side tubes, and two studs support a plywood stop-end. By adjusting the position of the tie tube, and the size of the studs, the length of the wall can be varied.

For the erection of these modular forms, after reinforcement tying is complete, it is normal practice to progressively erect both faces of the forms; soldiers, panels and tubes, together. This makes tie installation a quite easy one man operation. (Figure 4.152) Wall fittings and penetration forms are also progressively fitted and fixed in position.

If one wall formface is completely erected, and the ties and other soldiers done as a later operation, tie installation becomes a two person task. One on the inner face of the

form to hold the cone and tie rod; the other on the outside to insert the bolt.

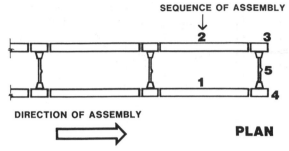

Figure 4.152 - WALL FORM ERECTION SEQUENCE

For stripping the wallforms, it is normal practice to completely dismantle the formwork. Nevertheless, it is possible to crane hoist one complete face of forms and tubes away from the concrete, after the removal of the bolts to the ties. However, when large metal formwork units are to be re-erected the alignment of the tie rod cones and the bolt holes can be difficult. The bolts used with these forms are too short to assist this alignment. They cannot be readily moved around as shown previously in Figure 4.44.

One of the major limitations of the all steel formwork systems is the weight of the panels and soldiers. Further, after a number of uses many of the panels become dented and this is reflected in the concrete face. To answer both of these criticisms plywood faced systems were developed.

As the cross section of the junction of two plywood faced panels in Figure 4.153 shows, specially shaped metal edges are rebated for the plywood face. The plywood (usually 12 mm) is screw fixed or rivetted to the frame and can be readily replaced if defaced. As the plywood/steel panels are generally lighter than all-steel panels they are made in taller sizes, up to 2400 mm.

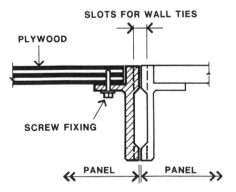

Figure 4.153 - JUNCTION OF PLYWOOD FACED PANELS

A plan of a formwork assembly for a wall corner is shown in Figure 4.154. The external and internal corner angles and narrow wall panels are the same as the panels of the all-steel system. The edges of all panels are slotted for connection devices, similar to the slots in the all-steel systems. As the inset details shows, these slots are also used for the fixing of the ties. At these positions a groove is provided on both abutting panels for access for a through-tie (snap-tie). This type of tie was discussed previously. (Figure 4.29) The tie and the panel connection is anchored by a double wedge arrangement as shown.

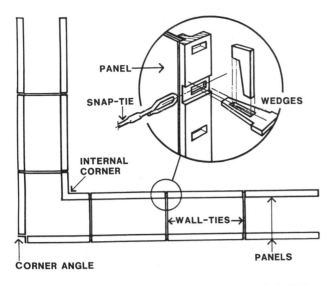

Figure 4.154 - ASSEMBLY OF PLYWOOD FACED PANELS

Rows of ties are provided along the vertical edges of all wall panels and, with this type of snap-tie, the formwork must be totally dismantled for stripping.

Wall alignment is achieved with horizontal tubes as for the steel panels. If the form consists of two or more rows of panels, vertical tubes are required for vertical alignment as shown in the vertical section of a single formface in Figure 4.155. Bracing is similar to that for the steel panel systems.

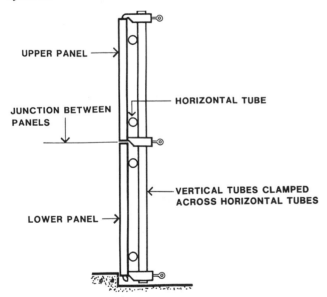

Figure 4.155 - VERTICAL TUBES FOR ALIGNMENT

In general, both the steel and plywood faced modular systems are unsuitable for concrete work which has a high visual quality requirement. Moisture loss occurs at most panel to panel junctions and grout loss can also occur there and at the ties. The dismantling that is usually needed at stripping makes them unsuitable for the repetitive production of similar shapes. Further, the modularity of the systems limits applicability to simple shapes.

However, for simple one-off shapes these proprietary systems can have cost and labour advantages. Studies have shown that the erection of modular proprietary systems

often takes less work time than the equivalent conventional wall form. For simple shapes it can be as little as half as much. Also, the labour required need not be carpenters, experienced labourers can handle most aspects of the work.

Proprietary Components for Large Wall Forms.

The more extensive use of cranes on site has encouraged the use of larger wall forms. To match this, a range of proprietary products for larger wall forms are available.

In most cases they are devised to work with the other proprietary products, panels, ties, corner units etc, to provide a complete wall formwork system. Additionally, they are often used in conjunction with the conventional wall formwork components, of plywood and timber framing members, for large formwork assemblies. A great diversity of components is available. Only the most common ones will be described.

The most common products are heavy duty double channel units, adjustable and fixed cantilever bracket units, lightweight channel members for joists, screwjack ends for channel units, and hoisting attachments. Manufacturers' literature should be refered to for information on the full range of products.

The double channel unit, often referred to as a soldier, consists of two cold rolled channels, set back to back, with end plates that are drilled for bolting. They are available in a range of modular lengths and can be extended by bolting end-to-end. Figure 4.156 shows a vertical section through a wall form, using soldiers as vertical walers, lightweight metal channels as joists and plywood formfacing.

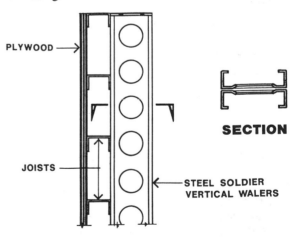

Figure 4.156 - STEEL SOLDIER AND CHANNEL FORM

The type of soldier shown has a series of flared penetrations through the web which, in addition to reducing the weight of the section, also provide access for horizontal stiffening members, and can be used for hoisting attachments. Figure 4.157 shows the fixing of the plywood form face to the lightweight channel joists.

The channel section is provided with a regular sequence of pairs of holes. The larger one provides access for the head of the power screwdriver. The smaller one, on the face to the back of the plywood, permits screw fixing of the joist to the plywood.

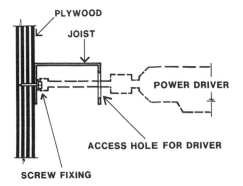

Figure 4.157 - CHANNEL FIXING TO PLYWOOD

These larger holes on the back leg of the joist are also used for bolting the joist to the soldier. Large washers are needed to suit the hole size and the space between the channels of the soldier. (Figure 4.158)

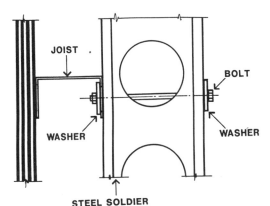

Figure 4.158 - CHANNEL FIXING TO SOLDIERS

Zed purlin sections can also be used as joists. Figure 4.159 shows the fixings to the plywood and the soldier.

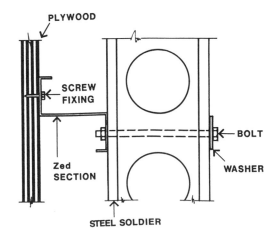

Figure 4.159 - Zed SECTIONS AS JOISTS

It should be noted that, while the plywood is always fixed to every joist, it is not always necessary to fix every joist to the soldiers. The high fluid concrete pressures that occur with deep wall forms usually result in the quite close spacing of the joists. As a result, between the top and bottom joists which are always fixed, usually only every second, and sometimes only every third, intermediate joist is fixed to all the soldiers. The

determination of the number to be fixed is based on the need for formwork rigidity, especially during hoisting. A matter as important as this should be calculated by an engineer competent in formwork design.

Timber joists are often used with steel soldiers. The simple fixing of bolting the joist to the soldier is shown in Figure 4.160.

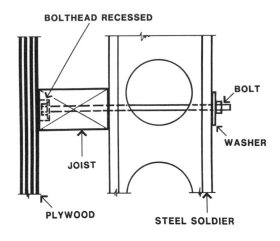

Figure 4.160 - TIMBER JOIST FIXING - 1

A variation of this that permits faster dismantling of the form is given in Figure 4.161.

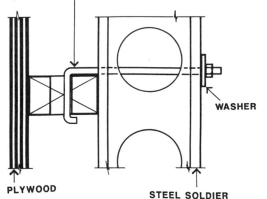

Figure 4.161 - TIMBER JOIST FIXING - 2

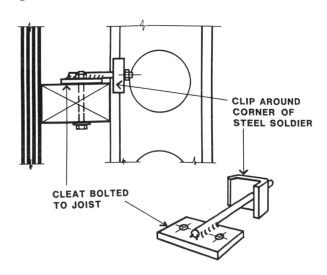

Figure 4.162 - TIMBER JOIST FIXING - 3

Figure 4.162 gives a detail of a clip fixing that also permits fast dismantling. Cleats, with threaded rods welded on, are bolted to the joists. Clips on the threaded rods hook around the lip of the steel soldier.

Where the soldiers span between widely spaced high capacity wall ties, it is often necessary to provide a horizontal stabilising member to prevent lateral buckling of the outer flange of the vertical soldier. This can be provided by bolting a rectangular hollow tube section to the outer flanges of all the soldiers as shown in Figure 4.163. Alternatively, universal beams or large timber sections can be used.

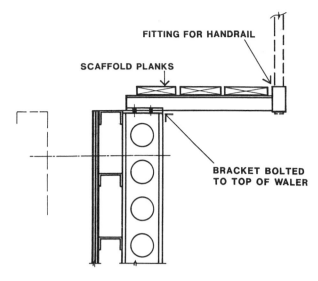

Figure 4.165 - CANTILEVER PLATFORM TO THE TOP.

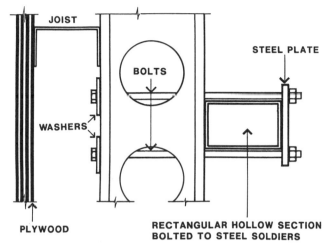

Figure 4.163 - RHS AS STABILISING BEAM TO SOLDIER

Some proprietary systems provide special fittings for the stabilising member to be located through the penetrations in the webs of the soldiers. (Figure 4.164)

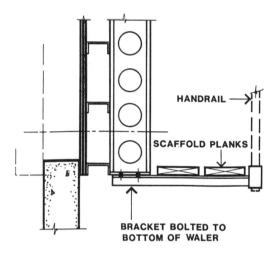

Figure 4.166 - CANTILEVER PLATFORM TO THE BOTTOM

These are suitable for worker access only, as they do not have sufficient strength to cater for heavy loads such as stacks of building materials. As shown in Figure 4.167 the typical brackets for much stronger and wider access platforms are provided with a support strut.

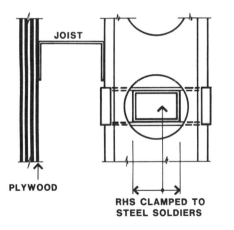

Figure 4.164 - INTERNAL STABILISING BEAM TO SOLDIER

The installation of stabilising beams through the holes in the webs of the soldiers is more awkward and time consuming than attaching a beam to the outer flange. Normally, it is only done with forms that are expected to have a large number of re-uses.

For the connection of access platforms to the vertical walers (soldiers) a range of standard components are available. Cantilever brackets for simple lightweight and narrow accessways can be bolted to the top (Figure 4.165) of the soldier or the bottom. (Figure 4.166) The bottom platform is used for access to the ties to the wall below.

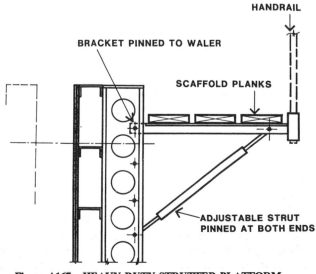

Figure 4.167 - HEAVY DUTY STRUTTED PLATFORM

The strut is usually adjustable in length, which permits the use of the brackets on sloping wall forms. With steeply sloping forms, such as that shown in Figure 4.168, the strut is not only adjusted in length, but is also connected to a hole located further down the soldier.

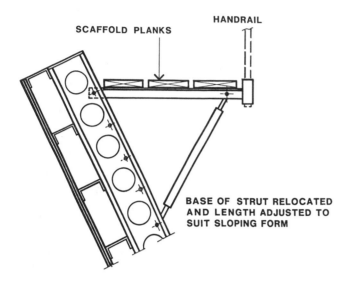

Figure 4.168 - ACCESS PLATFORM ON SLOPING FORM

This adjustable walkway bracket can also be used at the base of the wall form as a brace to adjust and control the plumb of the form. (Figure 4.169) The screw jack is fitted into the handrail bracket. This type of brace is not adequate for wind and impact loads, but is very useful during the formwork erection. To cater for wind and impact loads during erection, other braces or guys will be needed.

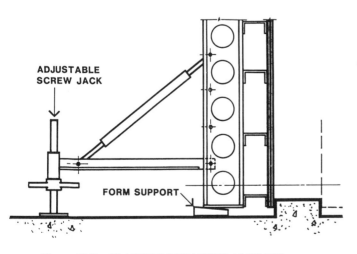

Figure 4.169 - PLATFORM BRACKET AS BRACE

The screw jack and the form in this case are shown seated on a concrete surface. Where the surface is soft, adequate soleplates will be needed to support both formwork and screw jacks.

The bolt holes in the end plates of the soldiers can be used for the attachment of lifting lugs. (Figure 4.170) These same loop fittings can also be fixed to both ends of a soldier so that it can be used as a cable spreader for formwork lifting. (Figure 4.171)

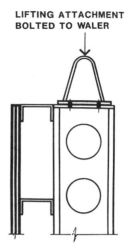

Figure 4.170 - LIFTING LUGS BOLTED TO SOLDIERS

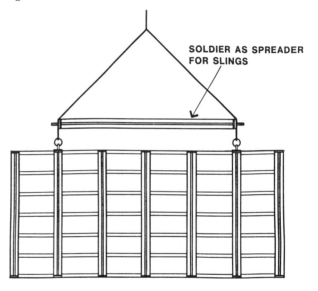

Figure 4.171 - SOLDIER USED AS A CABLE SPREADER

Another and frequent use for these soldiers is for the bracing of tall forms. As shown in Figure 4.172, the soldier provides a suitably stiff member for long bracing.

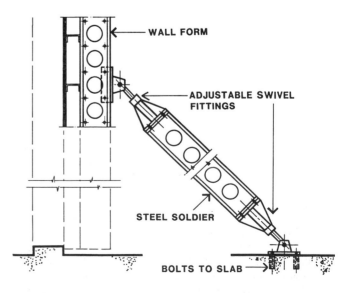

Figure 4.172 - SOLDIERS USED AS LONG BRACES

Adjustable swivel jointed screw jacks are bolted to the ends of the soldier. The eye fitting at the top end is pinned to the formwork soldier and the baseplate of the lower-end fitting is bolted to the base concrete or an anchor block.

Earlier in this chapter, Figures 4.04 and 4.14 illustrated the use the extended vertical waler in the alignment of wall formwork above a construction joint. Steel soldiers are very effective in achieving this.

As shown in Figure 4.173, the vertical walers are extended downwards by bolting another soldier section to them. At the base of this a lightweight access platform bracket and planks can fitted. Using tie rod locations from the previous pour, a proprietary fitting is installed on both sides of the wall to give control over the plumb alignment of the form assembly.

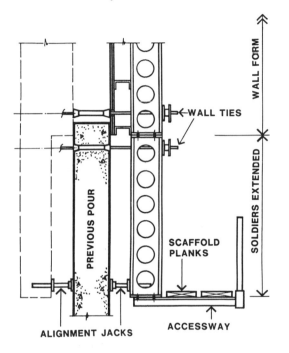

Figure 4.173 - SOLDIERS EXTENDED DOWNWARDS

In this example, the weight of the formwork is carried by the friction between the concrete face and the tightly clamped formwork. This requires that the formwork be carried on the crane until the adjusting and clamping is complete.

If a support cleat (shear bracket) is incorporated with the wall tie and clamped to the wall with the tie, the process is simplified. One face of the formwork can be hoisted and safely clamped, and then the other form can be hoisted.

He-bolts are a suitable tie for this. The cones on each end of the embedded tie rod effectively anchor it into the wall and permit a load to be applied at one end without the he-bolt being at the other end. Bar-ties are unsuitable for this, unless the support cleats are arranged to permit the fitting of additional washers and wingnuts at both faces of the concrete wall. Coil ties are usually unsuitable due to their low load capacity.

In the discussion so far, the emphasis has been on heavy duty steel proprietary components. Aluminium framing members which are suitable for joists and vertical walers are also available. Figure 4.174 shows the cross section of two of the ones which are available.

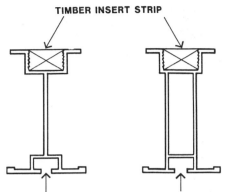

Figure 4.174 - PROPRIETARY ALUMINIUM BEAMS

There are two essential features of these aluminium beams. Firstly, the top has a timber strip installed in a groove formed in the aluminium extrusion. This enables the nail or screw fixing of the formface plywood. There are no direct connections, nails or screws, retaining the timber strip in place. It is cut accurately to size, and is retained in place by the serrations formed in the aluminium extrusion.

Secondly, the lower flange is provided with a continuous recessed groove for the easy installation of 'tee' headed bolts. Where the aluminium beams are used as joists, as illustrated in vertical section in Figure 4.175, the wall formface is nailed to the outer (top) flange and the double channel steel soldier is fixed with 'tee' headed bolts and clips to the inner (lower) flange.

As will be described in Chapter 6, these aluminium beams are also widely used in soffit formwork.

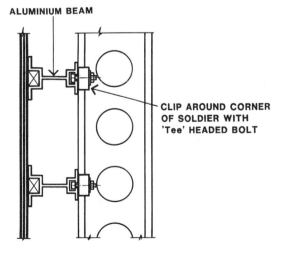

Figure 4.175 - ALUMINIUM BEAMS AS JOISTS

Figure 4.176 shows the vertical section of a wall form where the aluminium beams are used as horizontal walers with the studs being timber. For ease of installation of the wall ties, the aluminium beams are used in pairs and a heavy washer bridges across them.

There is no access for the nailing of the timber studs to the beams so they are fixed to the lower flanges with clips and coach screws into the timbers. This enables the easy removal of the walers at stripping.

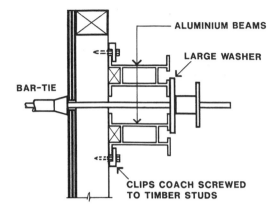

Figure 4.176 - ALUMINIUM BEAMS AS WALERS

CURVED WALL FORMWORK

Concrete walls that are curved in plan have long been common in civil engineering construction. Circular water tanks and waste water treatment plants are typical examples. In building work, the demand for this type of formwork varies with the ever changing architectural trends. Typical examples are walls to stairwells and curved building facades.

There are two main aspects to be considered in the arrangement of the formwork construction method: firstly, the selection of a suitable plywood thickness with respect to face grain direction and the radius of curvature, and secondly, the adoption of a formwork framing method appropriate to the curvature control requirements.

The table setout immediately below gives recommended bending radii for two levels of surface finish. For high quality concrete surfaces the second part of the table for a 'Low Reject Rate' should be used.

SAFE MINIMUM BENDING RADII FOR PLYWOOD WITH A MOISTURE CONTENT OF 10% AT ROOM TEMPERATURE WITH TWO DIFFERENT SAFETY FACTORS.

Refer to Figure 4.177 for definition of bending direction.

Plywood Thickness (mm)	Minimum Bending Radius for Expected Reject Rate of 5% (Safety Factor of 2)			
	HARDWOODS		SOFTWOODS	
	Bending Parallel with Grain (mm)	Bending Perpendicular to Grain (mm)	Bending Parallel with Grain (mm)	Bending Perpendicular to Grain (mm)
4	254	177	406	254
6	406	254	660	431
9	711	457	1117	762
12	1117	711	1778	1219

Plywood Thickness (mm)	Minimum Bending Radius if only Low Reject Rate can be Tolerated (Safety Factor of 3)			
	HARDWOODS		SOFTWOODS	
	Bending Parallel with Grain (mm)	Bending Perpendicular to Grain (mm)	Bending Parallel with Grain (mm)	Bending Perpendicular to Grain (mm)
4	381	266	609	381
6	609	381	990	584
9	1066	685	1676	1143
12	1676	1066	2667	1828

Note: Rejection is signified by face veneer cracking.

[Derived from graphs in "Bent Wood Members", United States Department of Agriculture, Forest Service, Forest Products Laboratory, Madison 5, Wisconsin No. R1903-3(1951)]

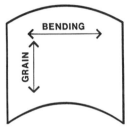

BENDING PERPENDICULAR TO GRAIN

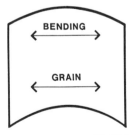

BENDING PARALLEL TO GRAIN

Figure 4.177 - DEFINITION OF BENDING DIRECTION

For large radius work, it is often convenient to form the wall as a series of segments, each of which has been fabricated to the appropriate curvature. Figure 4.178 shows a sectional plan of one example of this type of formwork.

The wallform is a vertical waler system, with waler sets and their wall ties located at the ends of each segment. As with straight walls that have vertical walers, which can be extended below the form, this method of forming is suitable for repetitive lifts of concrete pours.

As shown, the wall form segments are framed of horizontal joists, cut to the required curvature, with trimmer studs at each end. The joists span between the vertical walers at their ends.

In many cases the segments are sufficiently long for there to be other vertical walers sets at intermediate positions on the segments.

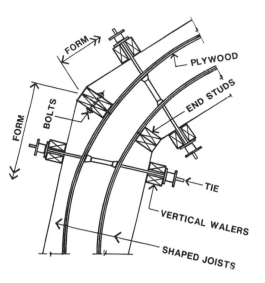

Figure 4.178 - LARGE RADIUS SEGMENTED FORM

The end trimmer studs of the outer segments are usually bolted together at the top and bottom. This aids the assembly and accurate alignment of the outer forms. Bolting is usually not necessary on the inner forms as they are contained at their base within the confines of the wall kicker and held by the ties.

The relationship of the wall ties to the formfaces should be noted. They are not at right angles to the surface. For this reason, bar ties, with their plastic cones, are often the most satisfactory tie. The plastic cones can accomodate some misalignment at the formwork suface.

With large radius work the junction of any two panels can be sprung inwards to relieve the closed circle of forms for stripping. For small radius work special provision for stripping must be made. At least one relief gap between segments should be used. Figure 4.179 shows this joint with a scarf cut in the plywood to aid stripping, and wedges clamped by bolting to control the gap.

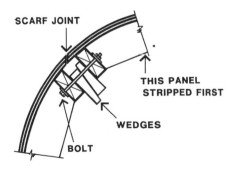

Figure 4.179 - RELIEF GAP FOR STRIPPING

Another method, that gives a wide stripping and adjustment gap, is shown in Figure 4.180. A curved metal cover strip (2 to 3 mm thick), with a continuous vertical slotted flat welded to it, bridges the gap. This slotted flat is similar to the edges of proprietary modular formwork panels.

The strip is held tightly to the formface with a proprietary clamp to a tube which bridges across the end studs to the segments. This tube can be bolted to the end studs to control the gap width.

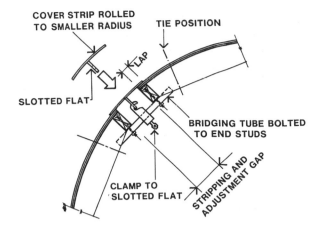

Figure 4.180 - COVER STRIP TO FORMWORK GAP

The sectional plan of the inner formwork of a vertical stud and horizontal waler formwork system is shown in Figure 4.181. The pairs of walers are fitted with curved packers to fix the correct position of the studs to the curve. At their junctions the walers are half lapped and bolted, with a packer, to give continuity. The construction of the outer formwork units is similar to that shown for the inner form.

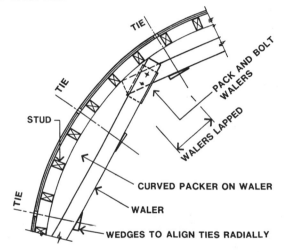

Figure 4.181 - CURVED HORIZONTAL WALER FORM

To give a square seating to the washers under wing nuts of the wall ties, wedges are provided. The form is assembled by erecting the flexible panels of plywood with the studs attached. These are seated on the base alongside the kicker. Extensive temporary bracing is needed to hold the very flexible panel to line. The walers are then added to the studs and the ties installed.

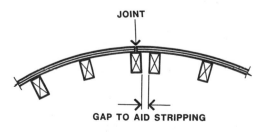

Figure 4.182 - JUNCTION AT THE FORMFACE

The junction between plywood panels is shown in Figure 4.182. To aid stripping, a small gap should be provided between the adjacent studs at these junctions.

This type of wall form is suitable for single lift work in a wide range of radii. It is not generally suitable for multiple lift work due to the lack of means to hold the flexible formfaces to line on the second and subsequent lifts. If fabricated in units consisting of formface, studs and walers, it requires additional framing to support it to line for further lifts.

A development of this method is shown in sectional plan in Figure 4.183. Only the inner form is shown as the outer form is similar in detail. Here the curvature is maintained by steel tubes rolled to the appropriate radius. The tubes are lapped at least three stud spacings at their junctions to maintain the accuracy of the curve. (Figure 4.184)

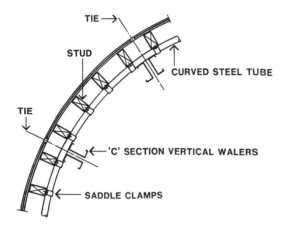

Figure 4.183 - STEEL TUBES AS WALERS

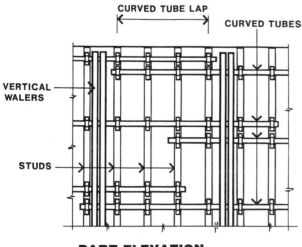

PART ELEVATION

Figure 4.184 - LAPPING OF CURVED TUBES

The vertical walers are shown as double steel channels. These could be two cold rolled 'C' sections or proprietary soldiers. Both of these are superior to timber walers in this case. Because of the round shape of the tube at the contact between it and the waler, and the large loads on walers, there would be crushing of the faces of timber walers if they were used. (Figure 4.185)

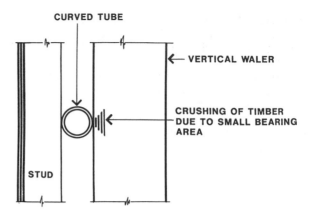

Figure 4.185 - CRUSHING OF A TIMBER WALER

A similar concentration of stress occurs at the interface between the tubes and the timber studs. However, the loads are much less and the crushing effect is usually not significant.

For work where there are to be successive vertical lifts, the proprietary steel soldier would be the best selection of the two steel types, due to the ease with which it can be extended downwards by bolting on another section.

One disadvantage of this method comes from the limited strength of the curved tubes. They are spanning between the walers and carrying the loads on the studs to the vertical walers. This normally results in the vertical walers being quite closely spaced with consequent increase in costs.

The tube method is also used in most of the proprietary metal panel methods of forming curved walls. Figure 4.186 shows a sectional plan of a typical example of both faces of these forming systems. The basic unit is a flexible sheetmetal panel fitted with vertical slotted stiffener ribs at the edges and regular intermediate positions.

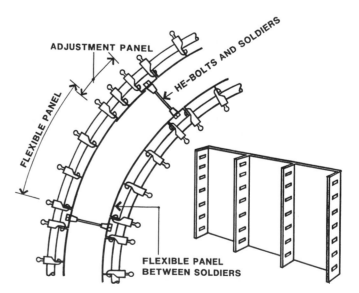

Figure 4.186 - PROPRIETARY MODULAR CURVED FORMS

Clamps, the same as those shown in Figure 4.145, are used to hold the flexible panel to the line of the series of curved steel tubes. Successive tubes can be lapped side-by-side at their junctions or have another tube lapped on top similar to that shown in Figure 4.146. Following the principles of the straight formwork systems as shown in Figures 4.142 and 4.143, the opposing faces of the forms are tied to each with he-bolts located on the centre lines of soldiers.

Like all proprietary panel systems these are modular products, and there is the problem of adjusting the length of both the inner and outer faces of the forms to the required perimeter lengths. For each adjustment a flexible panel with a single central vertical slotted stiffener rib is used. This is the same as that shown with timber forms as shown in Figure 4.180.

If there is perimeter length adjustment needed on the inner face then one or more of the distances between the soldiers can be extended with an adjustment panel. For the outer form an adjustment is required between every pair of soldiers to maintain the lines of the wall ties on radial positions. The total perimeter of the outer concrete face is greater than that of the inner concrete face by 6.28 times the wall thickness, 1256 mm for a 200 mm wall, 1845 mm for a 300 mm wall.

The techniques and some of the components of the proprietary system can be adapted for use with plywood faced forms. Figure 4.187 shows the sectional plan of the inner form of one such method.

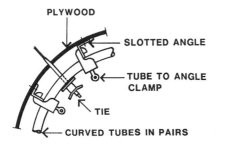

Figure 4.187 - PROPRIETARY ITEMS WITH PLYWOOD

Proprietary slotted vertical angles are screw fixed to the curved plywood formface at regular intervals. Pairs of curved tubes, acting as horizontal walers are clamped to the slotted angles. The lapping of these pairs of tubes should be done similar to the waler lapping shown in Figure 4.13. Large washers under the wingnuts of the wall ties, bridge across the pairs of tubes. He-bolts or bar ties are the most suitable as they maintain the wall thickness.

As the pairs of tubes are often quite closely spaced there is a correspondingly large number of wall ties. It is possible to reduce the number of ties by adding a pair of vertical walers on this tie line. Timber walers should not be used due to their tendency to crush over the round tubes with the large loads acting on the walers.

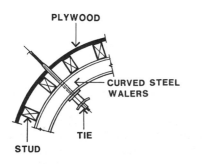

Figure 4.188 - CURVED 'I' BEAM WALERS

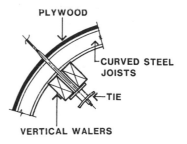

Figure 4.189 - CURVED 'I' BEAM JOISTS

Figures 4.188 and 4.189 show the sectional plans of the inner face of curved formwork with curved steel members and vertical timbers. The curved steel members used are usually small I beam sections. For the horizontal waler form, the studs are usually screw fixed to the walers which are in pairs to facilitate easy tie installation and effective lapping.

The vertical waler form is the most suitable for the work with successive vertical lifts, and the end of each curved unit is usually fitted with a vertical end stud to the joists, to permit bolting to the adjacent form unit.

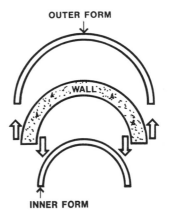

Figure 4.190 - SMALL CURVE STRIPPING PROBLEMS

Very small radius work, such as the curved ends of stair wells, can pose a stripping problem if the shape is a full semi-circle and the formwork faces are constructed in single units. Figure 4.190 illustrates the problem. Both the inner and outer formfaces, at the ends of the semi-circle, are effectively parallel to the direction of stripping. As a result, the start of the stripping will involve a sliding action between the formfaces and the concrete. There is the strong possibility of both formface and concrete damage. Where the concrete face is specified to be high quality concrete this situation is unacceptable.

Figures 4.191 and 4.192 show the sectional plans of the outer and inner forms of a system that caters for this problem. Both forms have vertical studs fixed to curved steel members; in this case steel angles are shown. The steel members are located at the top and bottom of the form and at intermediate spacings related to wall tie and waler strength. The vertical walers are not shown on the details.

In both cases, the angles have pairs of extended arms which are connected to a rod fitted with a turnbuckle. At the start of stripping the turnbuckles are tightened to flex the outer form away from the concrete face and the inner form inwards from the concrete surface. Both forms can then be carefully stripped out without abrasion between formfaces and concrete.

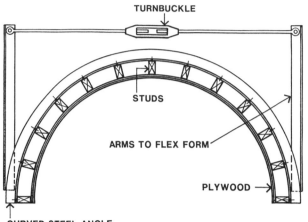

Figure 4.191 - OUTER SEMI-CIRCULAR FORM

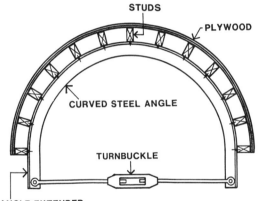

Figure 4.192 - INNER SEMI-CIRCULAR FORM

The use of turnbuckles to change the curvature of a formface is not confined to stripping. One proprietary curved form system uses turnbuckles to vary the curvature. Figure 4.193 shows the plan view of one face of this system.

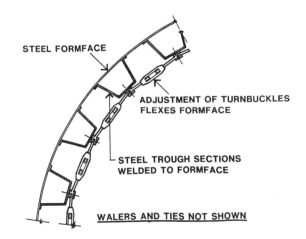

Figure 4.193 - PROPRIETARY FLEXIBLE FORM

The formwork comprises a flexible steel formface which has vertical steel trough sections (of folded sheetmetal) welded to it. The inner flanges of the troughs are connected by lines of turnbuckles at the top and bottom. Adjustment of the turnbuckles changes the curvature. This can be to a range of radii or to a composite curve. The cellular shape gives the formface great flexural strength which permits wide spacing between the lines of horizontal walers and wall ties.

Extreme accuracy and adherence to a truly curved shape is not always required. For many large radius curves it is often possible to form the concrete face with flat form units such as proprietary plywood and metal framed formwork panels. Figure 4.194 shows a sectional plan of this type of formwork.

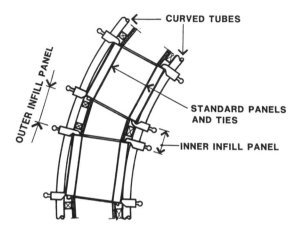

Figure 4.194 - CURVES FROM STRAIGHT PANELS

Both faces consist of a continuous sequence of standard panels and infill panels with through ties (snap ties) on all panel junction lines. The external and infill panels are different in width to each other to accomodate the different external and internal perimeters.

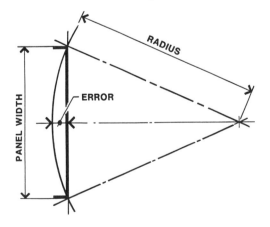

Figure 4.195 - ERROR BETWEEN CIRCLE AND CHORDS

Figure 4.195 illustrates the error that occurs between the true circle and the series of chords to the circle that the flat panels form. For a circular tank of 20 metres internal diameter the error would be 4.5 mm on a 600 mm wide flat panel.

This technique can also be used for purpose-built forms. Figure 4.196 shows a sectional plan of the panel arrangement. A sequence of simple flat plywood faced and timber framed panels are bolted to folded sheetmetal channels.

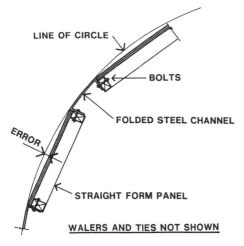

Figure 4.196 - STRAIGHT FORMS AND METAL PANELS

The difference between the inner and the outer perimeters is accounted for with a different width of inner to outer channels. The waler and tie system is not shown. It could be timber walers with packers similar to that shown in Figure 4.181 or curved tubes similar to Figure 4.183.

SINGLE FACED WALL FORMS

In the details of construction previously discussed for double faced formwork assemblies, the very high loads from the fluid concrete pressure were balanced through the tying of one formed face to the other by the wall ties. Therefore, the concrete pressures had no effect on the overall stability of the formwork assembly. In single faced forms no such balance occurs and the formwork must be braced and anchored to prevent instability and movement.

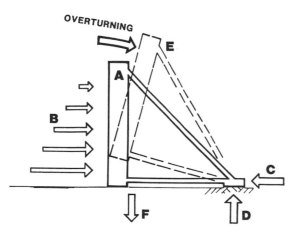

Figure 4.197 - SINGLE FACED FORM OVERTURNING

Figure 4.197 diagramatically illustrates the destabilising action of the fluid concrete pressure on a single faced formwork unit. The formwork assembly 'A' consists of a formface structure with a raking brace from the ground up to its top and a horizontal tie from the bottom of the raking brace to the bottom of the formface structure.

The fluid concrete pressure 'B', increasing with depth, acts horizontally. The total forces from this concrete pressure must be resisted by a suitable anchorage force 'C', to resist sliding. The form, as a whole, will tend to rotate about this anchorage at 'C' and thus lift up towards position 'E'. This must be resisted by a downwards acting anchorage force 'F'. For stability there will be an equal reacting force at 'D'.

Part of the anchorage force 'F' will be provided by friction between the formface and the fluid concrete. However, no authenticated data is currently available on the extent of this friction and so it cannot be taken into account in the design calculations. It is usual to design the anchorage at 'F' to carry the full uplift requirements.

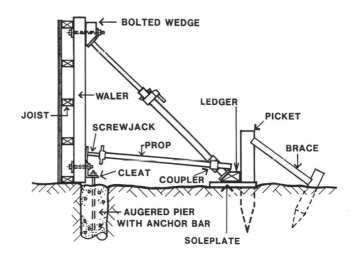

Figure 4.198 - CONVENTIONAL SINGLE FACED FORM

Figure 4.198 shows a simple conventional vertical waler timber framed form. The raking struts are proprietary telescopic props which bear on wedges bolted to the top of the form and onto a shaped ledger at the bottom. To cater for the large vertical and horizontal forces on the ledger, wide sole plates are used. Braced pickets at the back of the ledger cater for the horizontal forces.

In this case, the anchorage at the base of the wall is provided by an augered pier located beneath each vertical waler. A threaded rod is cast into the pier, to lap with the reinforcement, and is angle cleated to the walers. The anchorage effect of the augered pier is provided by the weight of its concrete and the adhesion between the ground and the concrete face of the pier.

If the surface were rock, the anchorage would be rock anchors set at a suitable depth. For anchorage to a concrete slab or footing, expanding anchors would be used.

For repetitious use, a fabricated steel assembly can be the most economical. Figure 4.199 shows the cross-section of an example. The fabrication into large units permits their relocation by crane.

The provisions shown for hold down anchorage are similar to that used for the simple form shown in Figure 4.198. For the vertical forces at the base of the raking brace a concrete bearing pad is provided.

Horizontal anchorage is usually provided by propping back to other construction work. This could be the base of columns, the face of excavations or retaining walls. In the absence of these, an anchorage similar to that shown in Figure 4.198 would have to be provided.

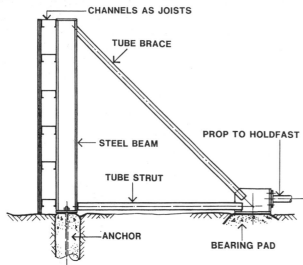

Figure 4.199 - PREFABRICATED SINGLE FACED FORM

The uplift action shown above can be reversed if the raking brace is provided as a tension tie on the concrete side of the form. Figure 4.200 shows the action of the forces diagramatically. The formwork unit 'A' has the fluid concrete pressures 'B' acting on it. These are resisted by a horizontal anchorage at 'C' and a raking angle tension tie that provides an anchorage force at 'D'. As a result of the action of the raking tie, the formwork unit 'A' will tend to move downwards. An upwards acting anchorage or support must be provided at 'F'.

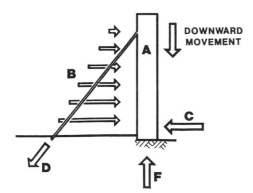

Figure 4.200 - SINGLE FACED FORM TIED BACK

Figure 4.201 shows an example of a simple single faced form anchored on the concrete side of the form. A raking augered pier provides the anchorage for the tension tie. This is an extended tie rod raking up to a she-bolt at the top of the vertical waler of the formwork. A proprietary angle bearing plate, which is bolted to the waler, provides end anchorage for the she-bolt.

A bearing strip provides the necessary support at the base. The horizontal anchorage is catered for, in this case by wedged strutting to an excavated earth or rock.

As with the previously described single faced form, the success of this method with be determined by adequate anchorage, the bearing strength under the wall form, and the security of the bank being relied on for horizontal anchorage. These matters can only be determined by an appropriate soils investigation.

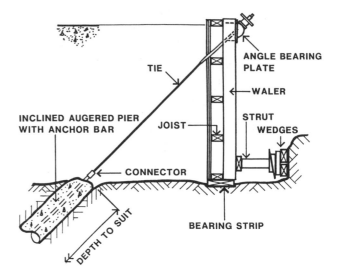

Figure 4.201 - EXAMPLE OF TIED BACK FORM

This type of single faced form can also be used where there is to be a vertical repetition of pours. (Figure 4.202) The raking ties are connected to pigtail anchors cast into the previous pours. The horizontal and vertical anchorage of the base of the form is provided by bolting the extended vertical walers to pigtail anchors also cast into the previous pour.

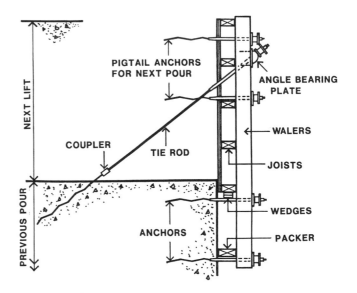

Figure 4.202 - CLIMBING SINGLE FACED FORM

Wedges between the underside of the form and the she-bolts only provide vertical adjustment. The resistance to vertical downward movement of the form, in this case, comes from the tight interface of concrete and the formface causing friction between form and concrete. For other than small forms, seating brackets may need to be bolted to the concrete face to provide vertical resistance to downward formwork movement.

For a sloping single faced form, such as that required for the face of a concrete dam, the resistance to the concrete pressure can be provided by stacking weights (kentledge) on the form. To carry these weights, the heavy duty walkway platforms shown in Figures 4.167 and 4.168 are usually suitable.

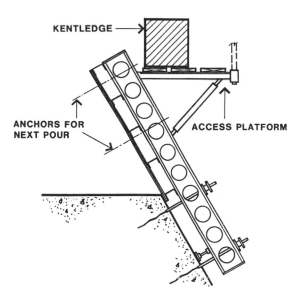

Figure 4.203 - SINGLE FACED SLOPING FORMWORK

Figure 4.203 shows a cross-section of this type of sloping formwork. The 'vertical' walers are anchored to the previous pour and cantilever to the top of the form. At each lift, anchors are cast in for use in the next pour.

Brackets built onto the walers support a platform for the weight of the kentledge. The vertical walers must be checked for the two bending situations: the walers carrying the kentledge and the walers resisting the concrete pressure with the assistance of the kentledge.

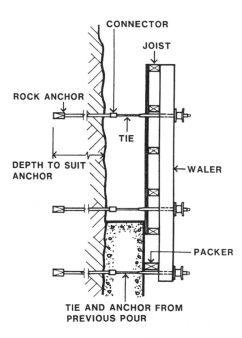

Figure 4.204 - TIED BACK TO A ROCK FACE

For narrow vertical pours, it is usually not possible to install the raking tie. If the excavated face consists of a rock suitable for the installation of rock anchors, these can be used to hold horizontal ties and resist the fluid concrete pressures. (Figure 4.204)

Here, the fluid concrete pressure is being resisted by horizontal ties anchored into the rock. There will be no downthrust from this tying method. The only vertical load

that must be carried is the weight of the formwork and the access platforms.

Similar to the double faced forms, single faced forms require attention to tying at the corners. Most of the techniques shown for double faced forms will work for these forms. However, with the greater thickness of these walls it is often possible to tie the corner through the concrete. This is shown in sectional plan in Figure 4.205.

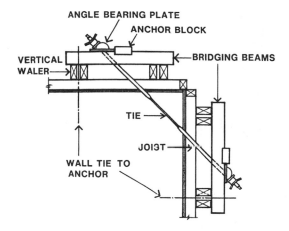

Figure 4.205 - CORNER TYING FOR SINGLE FACED FORMS

To provide fixing points for the ends of the ties, pairs of horizontal bridging beams span between and are fixed to two vertical waler pairs. Angled anchor blocks are checked into these bridging beams to prevent the angle bearing plates sliding.

CHAPTER 5: COLUMN FORMS

The function of column formwork is to enable the construction of columns that have the specified surface quality and are accceptably accurate in shape and position with good alignment to other adjacent columns, walls and building facades.

Because they usually have a small volume, column forms are filled rapidly, and full hydrostatic pressure from the fluid concrete occurs. As shown in Chapter 1, Figure 1.16, this pressure is further enhanced by the reflection of the vibration in members where all plan dimensions are less than 2 metres. This applies to most columns.

The other factor that affects the fluid concrete pressure is the discharge height of the concrete. (Chapter 4 Figure 4.05) Columns are often constructed with the column reinforcement extending well above the form. This is done so it can lap with the reinforcement of the next column or floor to be constructed above. Consequently, the concrete often has to be dropped from a point much higher than the top of the column form.

The impact effect of dropping the concrete can greatly increase the concrete pressure in the form. With fast pours, which are common in column work, the pressure can reach the hydrostatic pressure applicable to the full height of the column, and can act at all levels.

Also, most columns are constant in shape over their full height and can be, and often are, used either way up. Therefore, it is good practice to design them for the full discharge height hydrostatic pressure, over the whole form height.

As in all formwork activity, safety is the paramount consideration at all times. Safe access must be provided for the site workers and the several parts of the form must be stable and effectively braced at all stages of the work. This bracing not only has to hold the column form true to position and plumb, but must also contribute resistance to impact loads. These can occur when hoisted loads and concrete buckets collide with forms. This topic is covered in more detail later in this chapter.

The accuracy requirements for columns vary according to their position in the building. The tolerances, which are the maximum permitted deviations, are normally specified in the project documentation. These should cover plumb, face steps, twist and the relationship to absolute position.

Where the column is on the outside of a building each successive pour is required to be an extension of the previous one below. Here visual qualities are important. Figure 5.01 shows the elevation of a column built with alternating position deviations at successive floors. The resulting plumb deviations would be very evident when sighting up the column from below. Any twist in the column would further emphasise misalignments with the facade. The deviations from absolute position could cause further problems of fit with the facade.

Consequently, the tolerances for fully exposed external columns are quite stringent. In addition to the usual requirements covering plumb in both directions, twist and absolute position, tolerances can be given on distances between adjacent columns and the variations from plumb and straightness over a number of successive columns or even the whole height of the building.

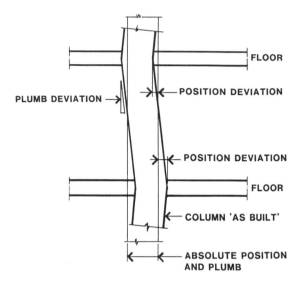

Figure 5.01 - DEVIATIONS IN EXTERNAL COLUMNS

For an internal column the accuracy requirements are not as stringent as those for external columns. Internal columns are not always positioned exactly over the top of the column below. The deviations that can occur with an internal column are shown in Figure 5.02.

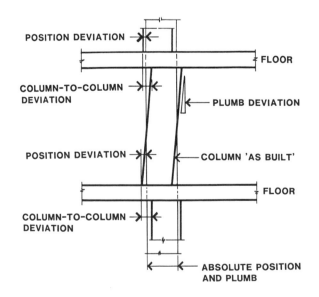

Figure 5.02 - DEVIATIONS IN INTERNAL COLUMNS

The permitted amount of plumb deviation will be limited by two concerns: structural and visual. The first one of structural considerations is usually the less stringent of the two. Excessive deviations can lead to eccentric moments acting on the column.

Visual concerns relate to the sighting of obvious misalignments between the column edges and the lines on other vertical elements nearby , e.g. door jambs, window jambs and other columns. The plumb tolerances in these cases may be quite small.

Position deviations also have structural and visual limitations. Column to column deviation will cause eccentric moments on the columns. Errors in position will be visually obvious when there is misalignment with modular ceilings. Alignment with the modular ceiling is also a limitation on the permitted degree of twist deviation and out of shape of the form.

Like all other formwork construction, three distinct stages apply. **STAGE 1** is the erection and completion of the forms ready for the pour. It includes formwork preparation, cleaning and applying release agent, hoisting in sub-assemblies or as a whole, the provision of temporary bracing and access platforms so that it is safe at all times and, finally, the completion of the bracing and accurate alignment of the formwork.

During **STAGE 2** the formwork is subject to the high pressures from the fluid concrete and the energy input of immersion vibrators. At this time the probability of impact from concrete buckets is greatest.

Throughout **STAGE 3** the formwork must support and protect the concrete until it has developed sufficient strength and surface hardness to be ready for the stripping of the formwork. If the column is to have high quality concrete surfaces, this stage will include the installation of surface protection so that the column will not be defaced by staining or impact.

The type of formwork and the method of stripping associated with it, can be a determinant of the concrete surface hardness required to resist surface damage during stripping. One example of this situation is the hinged form.

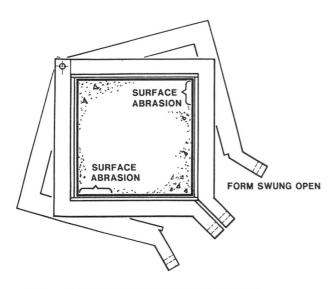

Figure 5.03 - HINGED COLUMN FORMWORK

As Figure 5.03 diagramatically indicates, at the commencement of the opening of the form there is abrasion at the column corners as the formface and the concrete face slide in relationship to one another. Hinged forms of this type should not be used where a high quality surface is specified. Details of a hinged formwork system that avoids this problem are given later in this chapter.

RECTANGULAR COLUMNS

As one of the more commonly used clamping devices, which is widely used for most types of column forms, originated in the development of proprietary modular column forms, these formwork systems will be covered first.

PROPRIETARY FORMS

The basic proprietary column form assembly is shown in the sectional plan of Figure 5.04. The standard steel framed, plywood or metal faced form panels span between the corners where they are connected to slotted corner angles with the screw clamps or double wedges used in wall forms. (Refer to Figures 4.144 and 4.154)

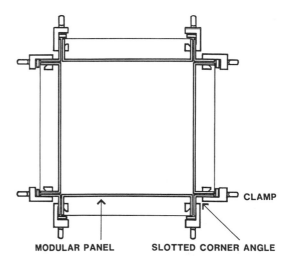

Figure 5.04 - PROPRIETARY COLUMN FORMWORK

Column clamps, of the type shown in Figure 5.05, fit around this assembly, bearing tightly on the corner angles and the panel edges. These provide a tensile capacity along all four column faces to resist the tendency of the fluid concrete pressure on the forms to burst them open at the corners. The clamps also aid in holding the corners of the formwork assembly square.

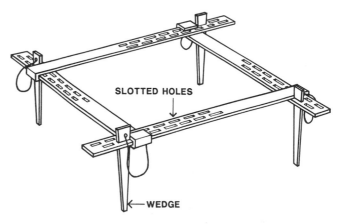

Figure 5.05 - COLUMN CLAMP

The bearing area between the narrow edge of the steel column clamps and the similarly narrow edges of the corner angles and panel edges is quite small and, therefore, the stresses at this interface are very high. (Figure 5.06) However, they are within the capacity of structural steels. It is the use of these clamps bearing on timber that causes a problem. This matter is discussed later in this chapter.

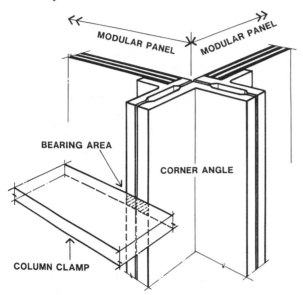

Figure 5.06 - BEARING AREA - CLAMP TO COLUMN

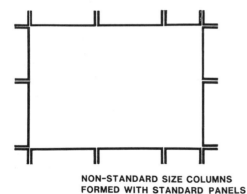

NON-STANDARD SIZE COLUMNS FORMED WITH STANDARD PANELS

Figure 5.07 - ADDING STANDARD PANELS

If the dimensions of the column to be formed are not those of available standard modular panels, two or more narrower standard panels can often be added together to make up the required formface width. (Figure 5.07)

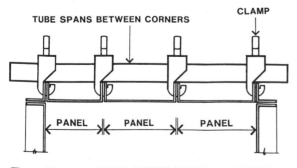

Figure 5.08 - TUBES TO CARRY LOADS TO CORNERS

A series of panels cannot span between the corners. To carry the pressure across the formface, horizontal tubes are

clamped across the corner angles. The part sectional plan of Figure 5.08 details one face of the column form. As shown, the panel edges bear on the tubes and the tubes carry the loads to the corners.

The column clamps and tubes are not designed to provide this bending capacity for other than small columns. If used for large columns, excessive deflection of the column face usually occurs.

PURPOSE MADE COLUMN FORMS

Conventional Timber Forms with Proprietary Clamps.

The steel column clamps described above are also used in conventional plywood and timber column forms. This method of forming small columns is shown in Figure 5.09.

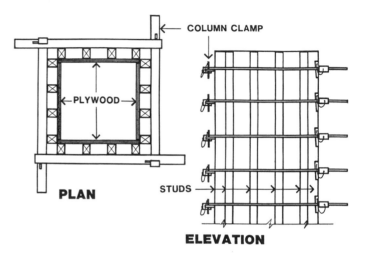

Figure 5.09 - CONVENTIONAL COLUMN FORMWORK

In the first instance the fluid concrete pressures are carried on the plywood which spans horizontally between the studs. In turn, the studs span vertically between the horizontal column clamps. At this stage an important point must be made on the selection of plywood for this case.

To avoid plywood joints, it is normal practice to cut the sheets lengthwise in widths to match the column faces. When used in this way the plywood is spanning structurally across the face grain. For most of the plywoods in use for formwork this is the weakest direction for resisting bending loads. However, plywoods with a thin face veneer have high cross-grain strength, and in some cases this transverse bending strength slightly exceeds the lengthwise strength. This was noted in Chapter 2. These plywoods should be used for this type of column form. Information on these plywoods is available from plywood manufacturers and marketing organisation.

The problem of the crushing of the timber at its contact with the narrow column clamps was refered to earlier in this chapter. When used with metal framed forms the permissible bearing stress, steel to steel, is 187.5 MPa for normal structural steels. When the bearing stresses act steel to timber, it is the crushing of the timber that is important even though the bearing area is larger than the steel to steel case. The maximum permissible bearing stresses for the softwoods usually used in formwork is approximately 5.0 MPa. In most cases some crushing of the timber face occurs at the interface with the steel column clamp. The location of this is illustrated in Figure 5.10.

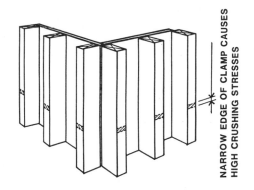

Figure 5.10 - CRUSHING OF THE COLUMN TIMBERS

Crushing results in movement outward of the formfaces under load. Grout is forced behind the corner fillets, and the end result is oversize columns with unsightly corners. (Figure 5.11)

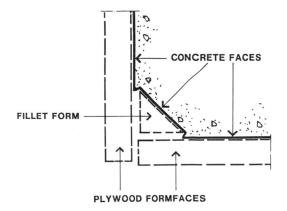

Figure 5.11 - CORNERS MOVEMENT FROM CRUSHING

It is important that this crushing be limited as much as possible. In addition to using the clamps at close spacings, in accordance with the manufacturer's directions, this crushing effect can be limited by attention to the details of the formwork framing. One method is illustrated in Figure 5.12.

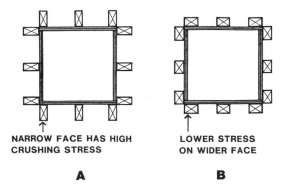

Figure 5.12 - WIDTH OF BEARING AREA

If the vertical studs have a narrow face, 'A', there will be high crushing stresses and considerable formface

movement. If studs with a wider face are used, 'B', the stresses will be proportionately lower and the crushing movement significantly less. Another column framing method that reduces the timber stresses at the interface with the narrow steel column clamps is shown in the part column form view of Figure 5.13.

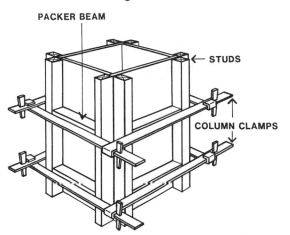

Figure 5.13 - ALTERNATIVE COLUMN FRAMING

In this case studs are used only at the corners and horizontal packer beams, of a depth that lines flush with the stud faces, are framed between the studs at each clamp position. The plywood spans two ways, horizontally between the studs and vertically between the packer beams. This method gives the greatest possible bearing area for the clamp and, therefore, the least crushing. Further, the packer beams lessen the bending action on the metal clamps and there is less formface deflection than that which occurs with the first framing method.

In both of the above methods, and indeed most column forming methods that use plywood formfaces, the corner junctions of the plywood must be detailed to achieve the maximum tightening action from the column clamps. Figure 5.14 illustrates this with part 'A' showing the correct method and 'B' showing a common fault that defeats that tightening action.

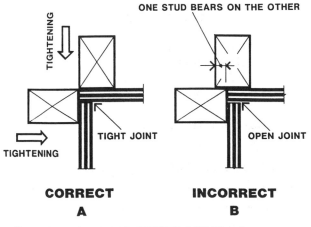

Figure 5.14 - PLYWOOD CORNER DETAIL

In part 'A' of Figure 5.14, the clamping action forces one side plywood face onto the other, and the inner face of a corner stud onto the end of the other plywood face. In part 'B', the lapping of one stud over the other prevents the effective tightening action between the two plywood sides. In the absence of a tight joint, grout loss and hydration staining results.

Inside the corner of the form the column corners are usually formed to have a 45 degree arris. As shown in Figure 5.11 above, any movement of the formfaces can result in a low quality corner. Even if formface movement is limited, the forming of this arris by the most common method, as shown in Figure 5.15, often leads to some loss of quality.

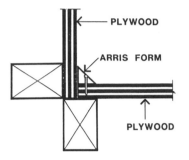

Figure 5.15 - TYPICAL ARRIS FORMWORK

The small timber or plastic fillet used is difficult to align accurately inside the assembled column form. If it is installed on the form faces before assembly there can be problems in preventing damage to the sharp outstanding edges during assembly. Secondly, it is not possible to get a tight interface of the fillet to the other formface.

For those cases where a high quality arris is needed, and the cost of a more complex form will be justified by a number of re-uses, the formwork shown in Figure 5.16 is suitable.

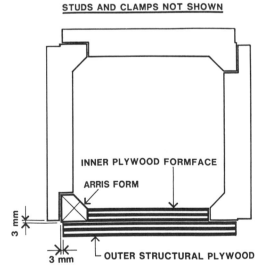

Figure 5.16 - HIGH QUALITY ARRIS FORMWORK

These arris forms do not have the vulnerable sharp edges of the previous case. However, to accomodate this bulkier section and control the alignment of the arris form, two layers of plywood are needed. The two layers of plywood can be screwed together from the back. Note the setback (usually 3 mm) of the ends of the outer plywood from the corners. This ensures that this outer plywood will not interfere with the clamping action at the corners. It is a method that also readily accomodates other arris shapes at the column corner, e.g. rounded, grooved, double filleted.

The fabrication techniques for these types of column forms are very similar to that described for wall forms in Chapter 4. Similar to wall forms, accuracy is important and care must be taken to avoid constructing a twist into the formfaces.

Erection of Conventional Column Forms

For formwork that uses the narrow steel column clamps, it is common to erect the forms one side at a time. Figure 5.17 shows one suitable sequence.

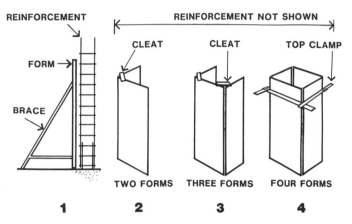

Figure 5.17 - ERECTION OF COLUMN FORMS

Formwork erection usually starts after the completion of the tying of the reinforcement and the installation of bar chairs on the reinforcement. However, sometimes one column formface is erected first, and braced in its final position to act as a guide for the accurate alignment of the reinforcement while it is being tied.

After the first side is secure in its position, the second side can be erected and secured by fastening the top corners together. Often only a single nail is used, but a safer method is to nail a plywood cleat across the top of the two adjacent corner studs.

The third and fourth column forms can then be erected in the same way. With the 'box' of the form complete, at least one column clamp should now be installed near the top. This will give the form assembly a good degree of safety. The installation of the remaining clamps and braces can then follow.

It will be obvious from this description that worker access to the top of the formwork will be needed throughout. A scaffold frame, complete with handrails, is needed for safe work access. Access platforms and bracing are topics covered in more detail later in this chapter.

Forms of this type can be fully assembled at the ground level and then hoisted over the reinforcement cage. (Figure 5.18) This is only possible if the reinforcement bars are vertical at the top and not bent over for lapping into the floor structure above.

Care must also be take to ensure that the formfaces do not slide on the reinforcement and damage the formface during the lowering of the form. One effective method is to temporarily place dressed vertical timber battens at the outer edges of the column ligatures. The column form can then slide down these battens and not collide with the reinforcement.The battens are withdrawn after the column form is in place.

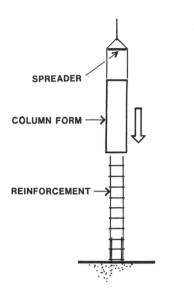

Figure 5.18 - LOWERING FORM OVER REINFORCEMENT

Stripping this type of form is simply a reverse of the piece by piece assembly. The top clamp should be the last one to be removed and this should not done until all four corners have been cleated together for safety. The sides of the column form can then be removed in sequence. Care must be taken at all times to ensure that any formfaces remaining in place are safe while awaiting removal.

Conventional Forms with Horizontal Walers

Column forms built with horizontal walers have many details that are similar to horizontal waler wall forms. Figure 5.19 shows a section plan of this type of formwork.

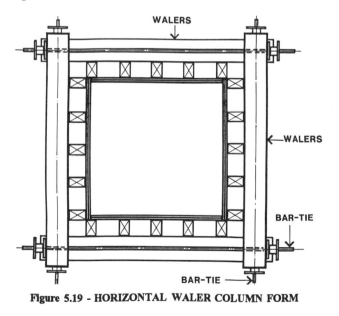

Figure 5.19 - HORIZONTAL WALER COLUMN FORM

Similar to the previous case the load path starts with the plywood spanning horizontally between the studs, which in turn carry the load vertically to the double walers. The walers then span between the ties which are located on the outside of the formwork.

This method of forming columns has a number of advantages. Firstly, the large bearing area at the interface between the studs and the walers results in low bearing stresses and minimal crushing. Secondly, the bar-ties

between opposite sets of walers can be very readily installed and removed. (Figure 5.20) If the walers are bolted in place (Figure 5.21) they remain on the forms at all times and the ties are the only items to be separated from the forms on stripping.

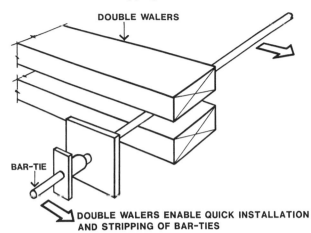

Figure 5.20 - EASY REMOVAL OF BAR TIES

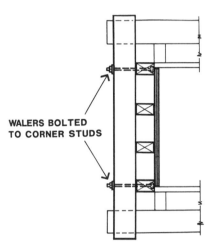

Figure 5.21 - WALERS BOLTED TO STUDS

With the walers bolted in place a further advantage can be the ability to bolt the bracing system permanently to the back of the walers. (Figure 5.22)

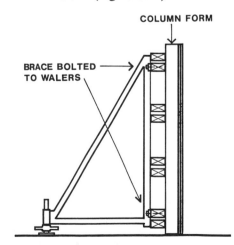

Figure 5.22 - BRACES BOLTED TO COLUMN FORM

Column form sides with the bracing attached are simpler and safer to erect than cases which have the bracing separate. The units are easy to hoist and are stable the moment they are placed in position. If, however, the braces are separate, the walers provide an easy means of safely attaching one form side to the other while being erected.

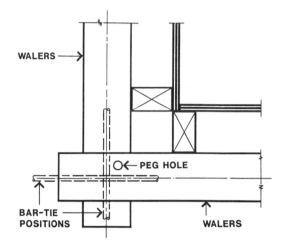

Figure 5.23 - TEMPORARY CORNER CONNECTION

As the part corner plan in Figure 5.23 shows, holes are drilled down through the pairs of walers where they cross at the corner. During column form erection a steel peg is placed in the holes to hold the adjacent loose form sides together. Note that the hole is inside the centreline of both the bar-ties to avoid conflict with these ties and not impede their installation or removal.

This pegging system also enables a safe procedure during the stripping of the individual column form sides.

Other Clamping Systems

It is not possible to cover the great variety of column tying and clamping systems that can be seen on construction sites. However, two more examples will be given. Both utilise prefabricated steel components made to suit the column size.

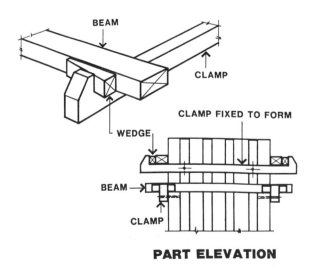

PART ELEVATION

Figure 5.24 - SPECIAL STEEL CLAMPS - 1

The first, shown in Figure 5.24, uses pairs of steel clamp members, with stop-ends, that are fixed to opposing form faces. These provide tension tying on the face they are fixed to. They also hold beams that are wedged across the ends of the clamp members. These loose cross beams, of steel or timber, carry the loads from the formface studs to the clamp members. This set effectively clamps one pair of opposing formfaces inwards. Immediately adjacent, another set clamps the other pair of formfaces at right angles.

A development of this is shown in Figure 5.25. By incorporating bar-ties into the equipment the hardware is simplified.

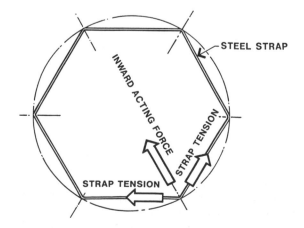

Figure 5.25 - SPECIAL STEEL CLAMPS - 2

The member with the stop-ends is made as a double waler. It provides both bending and tension capacity across the formface that it is fixed to. Bending strength across the other faces is provided by the loose beams that are wedged inside the stop ends, The tension capacity comes from the bar ties.

Both of these methods can be erected one side at a time or hoisted, completely assembled, over the reinforcement cage.

Perimeter Strapping

The strapping techniques that are common in the packaging industry can be effectively used to clamp column forms together. They are suitable for round, square and near square columns.

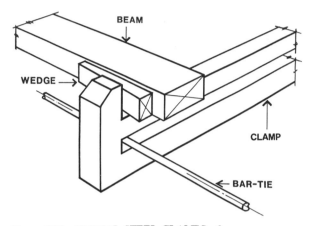

Figure 5.26 - CLAMPING FORCE FROM STRAP TENSION

The tie system consists of high tensile metal straps, metal anchoring buckles and a device to tighten the straps. For the hoop tension in the strap to effectively clamp the formfaces inward against the fluid concrete pressures, a near to circular shape in the strap is required.

The plan view of Figure 5.26 shows how the resolution of the forces along the strap line provides the inward acting forces. To maintain this 'hoop' action for rectangular columns, packers on the studs in the middle of the column sides, or deeper studs, are needed. Figure 5.27 shows the sectional plan of this type of column form.

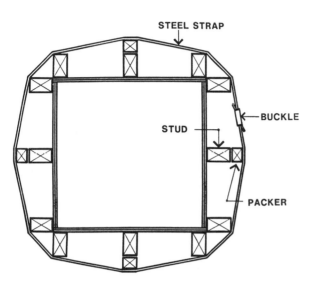

Figure 5.27 - NEAR-CIRCULAR COLUMN STRAPPING

If the studs in the middle of the formfaces do not have any packing, the straps cannot provide any inward acting restraint until the plywood deflects outwards. (Figure 5.28) The loads are then jointly resisted by the deflected plywood and the tension force in the strap.

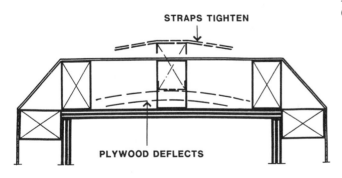

Figure 5.28 - DEFLECTION OF FORMFACE

To determine the number of ties that are required for a particular size and height of column form, reference should be made to the technical literature from the strapping manufacturer. These are not a high capacity tying system, and the spacing can often be quite close.

Stripping is quite simple; the ties are merely cut with tin snips. Two considerations arise here. Firstly, the high tensile steel straps are very springy and eye protection must be worn. Secondly, the straps are not re-useable and a large amount of rubbish is generated.

Two Part Column Forms

The principal of the two part column form is shown in the sectional plan of Figure 5.29. In a similar way, but to a lesser extent than that shown for hinged forms in Figure 5.03, the stripping of the form involves a relative sliding movement of the form on the concrete face at the start of the stripping. For this reason they are not normally suitable for columns that are to have high quality concrete surfaces.

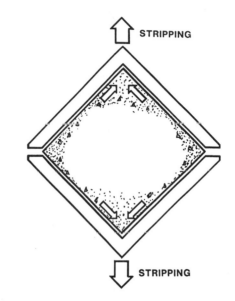

Figure 5.29 - STRIPPING A TWO PART COLUMN FORM

Where the columns are to be formed with vertical grooves or have bolts protruding from the concrete face, satisfactory stripping can be impossible.

For columns of 450 mm sides and larger these forms are usually constructed with plywood formfaces, vertical timber studs and prefabricated horizontal steel walers. Figure 5.30 shows the sectional plan of this type of column form.

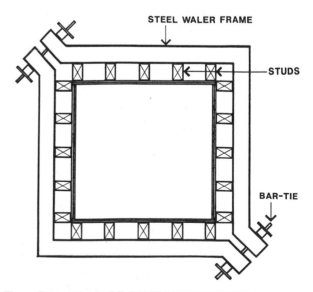

Figure 5.30 - TWO PART COLUMN FORMWORK

The walers can be single rectangular hollow sections, double steel channels or universal beams. The first is the most popular. Clamping is usually done with short lengths of bar-tie at the ends of the walers. This framing method can permit relatively wide spacing of the walers. Column forms 3 metres tall may only have five levels of walers. With only ten ties to undo, the stripping operation can be quite fast.

The right-angled shape of each of the two parts makes them reasonably stable during erection and temporary erection bracing is usually only needed in high wind locations.

For smaller column sizes the two part form can simply be folded from sheet metal, and the two parts bolted along their edges. (Figure 5.31) However, the required steel thickness may result in a formwork weight which is too great for man-handling.

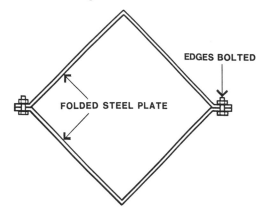

Figure 5.31 - FOLDED SHEET STEEL TWO PART FORM

A lighter form results if the sheet metal is fabricated with horizontal stiffening ribs at vertical spacings equal to half the column size or less, and vertical stiffeners at the middle of the sides. These forms are shown in Figure 5.32.

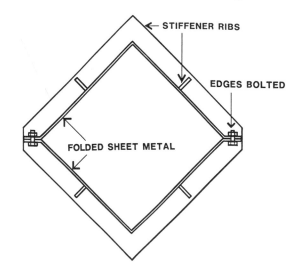

Figure 5.32 - STEEL COLUMN FORM WITH STIFFENERS

However, fabricating the steel form with stiffeners involves welding. In most cases the minor surface distortion that occurs with welding will be reflected in the concrete surface. This usually makes this type of form unsuitable for high quality concrete surfaces.

Similar to the first case these forms are shown with the edges bolted. The connection and later dismantling of column forms with bolted edges is time consuming. The

part sectional plan of Figure 5.33 shows a faster connection detail.

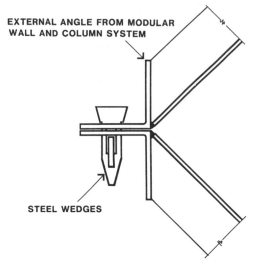

Figure 5.33 - FORM CONNECTION WITH WEDGES

The edges of the welded steel form are fabricated with the slotted proprietary external corner angles used in modular proprietary wall and column forms. These connect together with the double wedge assembly described in Chapter 4. Another method of speeding up assembly and stripping is shown in the sectional plan of Figure 5.34.

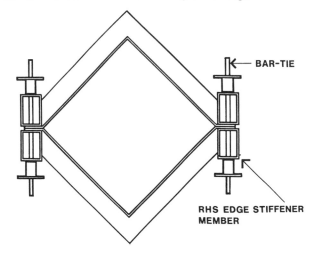

Figure 5.34 - EDGE CONNECTION WITH BAR TIES

Here the simplification is in the reduction in the number of fasteners. The fabrication of the column form with a stiffened edge, in this case shown as a rectangular hollow section (RHS), can reduce the edge fastening to a small number of high strength bar-ties.

Hinged Forms

A logical development of the two part form is the single hinged form. It has ease of erection and stability when both open and closed. As shown in the plan of Figure 5.35, it can be closed and stripped open, at columns on the edge of a suspended slab, with less than the usual hazards to the formworker. Access is only needed to the corner with the ties. By locating this corner as shown, the work area is kept away from the edge of the floor.

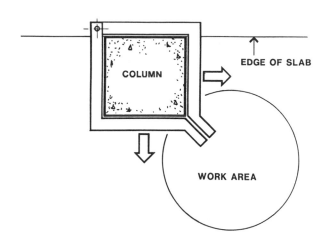

Figure 5.35 - SAFE ACCESS TO A HINGED FORM

A typical sectional plan of a hinged form is shown in Figure 5.36. Its construction is similar in many ways to the two part form of Figure 5.30.

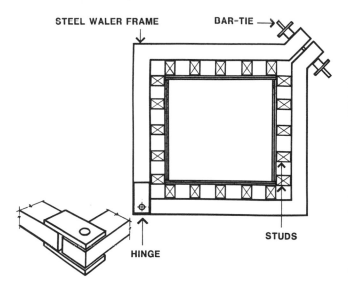

Figure 5.36 - HINGED COLUMN FORMWORK

In this detail the corner ties are shown as bar ties. Figure 5.37 shows a corner tie arrangement that is permanently attached to the walers. These hinged anchors are a 'quick strip' type enabling fast formwork removal.

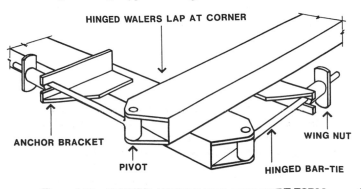

Figure 5.37 - CORNER ANCHOR FOR TWO PART FORM

The problems of concrete surface abrasion at corners was illustrated in Figure 5.03. If a more sophisticated hinge is used this problem can be avoided. Figure 5.38 gives plan details of the principles of an eccentric hinge pin.

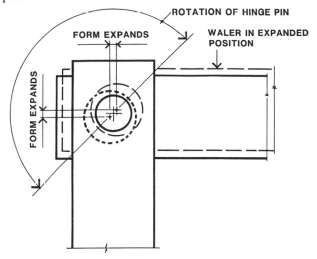

Figure 5.38 - ECCENTRIC HINGE TO AID STRIPPING

The vertical, full height, hinge pin connects all waler hinges together. At each waler hinge, one leg of the waler has a concentric bearing and the other has a bearing eccentric to the first bearing. The top of the vertical hinge pin is fitted with a locking arm. As a first operation in stripping the locking arm is released and rotated through 180 degrees. This effectively expands the forms by 3 mm or more in both directions. As a result of this increase in form size, when the other corner is unlocked and the forms are swung open, surface abrasion can be avoided.

Forms can be made with two or three hinge points and these offer advantages for particular cases. They can be used with grooved face columns and are often useful for columns that have protruding bolts. The sectional plan of Figure 5.39 shows the general principle of a two hinged column formwork. The hinges connect three sides together with one formface separate.

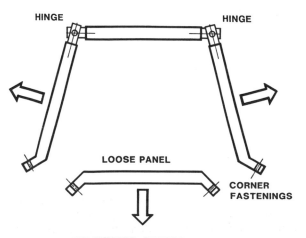

Figure 5.39 - TWO HINGED COLUMN FORM

The principle of the three hinge form is illustrated in the sectional plan of Figure 5.40. In both these cases the hinges and the corner fastenings can be fabricated similar to those used on the single hinge form.

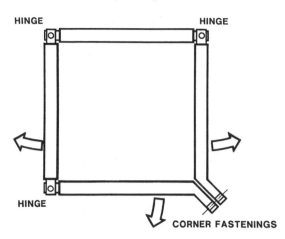

Figure 5.40 - THREE HINGED COLUMN FORM

Maintaining a right-angled shape, calls for braces on at least two hinges, at both the bottom and the top of the form. If the forms are used for external columns, in a situation similar to that shown in Figure 5.01, then the braces should be adjustable in length so any accumulated twist can be corrected. (Figure 5.41) This is an advantage that this system has over the rigid two part, single hinge types of forms.

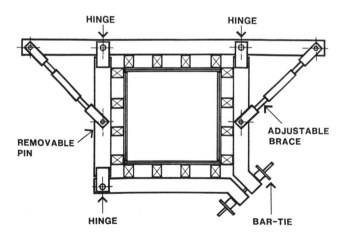

Figure 5.41 - BRACING TO HINGED COLUMN FORM

To achieve accuracy in the adjustment of the braces, turnbuckles with at least one lock nut are the most effective device. The braces must have a removable pin or bolt at one end so that they can be swung out of the way during formwork stripping.

Larger Rectangular Columns.

The methods described previously, where all the resistance to concrete pressures is provided around the outside of the form structure, are usually limited to columns that have maximum plan dimensions of one metre. For larger columns the horizontal timber walers usually become very large.

For smaller waler sizes, the horizontal timber walers, as shown previously in Figure 5.19, can be replaced by steel members. Steel walers, for larger column sizes, are

a satisfactory solution up to the tensile capacity of the external ties.

Beyond this, internal ties can be fitted. All the previously described framing methods, other than the narrow proprietary column clamps, can be easily fitted with internal ties. Figure 5.42 shows them installed in conventional timber horizontal waler forms with one line of internal ties for each direction. To avoid conflict where they intersect on the column centre line, there must be a small difference in the level of the ties. The gap between the walers, normally 30 mm, should be increased to 40 mm to accomodate this.

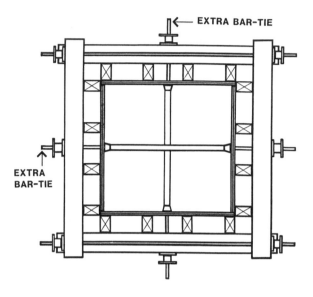

Figure 5.42 - INTERNAL TIES TO A LARGE COLUMN

Figure 5.42 shows the ties on the centreline. However, if there is reinforcement on the centreline, the ties must be offset or, for a symmetrical appearance of the cone holes, a pairs of ties on each column side will be needed.

Internal ties can also be fitted to hinged forms. Figure 5.43 shows a sectional plan of this for a single hinged forms.

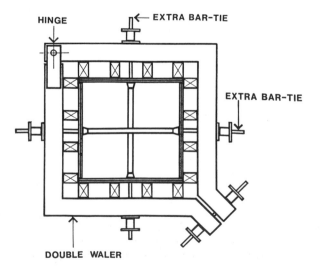

Figure 5.43 - INTERNAL TIES TO A HINGED FORM

As shown the walers are made as pairs for the symmetrical seating of the washers to the ties. Alternatively, single horizontal walers can be used if they are wide rectangular hollow sections. These walers are drilled for the ties.

A method that does not dictate that the ties must all be at the same levels as the walers is shown in Figure 5.44. Outer vertical double walers are installed centrally. The ties can be positioned anywhere on these walers.

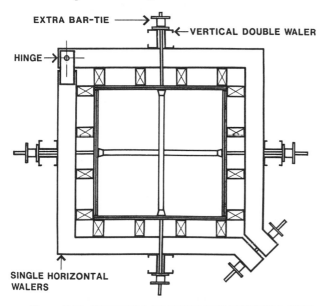

Figure 5.44 - VERTICAL WALERS FOR INTERNAL TIES

CIRCULAR COLUMNS

Circular columns can be formed by a number of methods, using proprietary products or being purpose made from normal construction materials. Following the order of first part of this chapter, proprietary forming methods will be covered first.

Three materials are used for spirally wound, cylindrical forms: cardboard, plastic and steel. In general, these proprietary products are single-use disposable column formwork systems.

Cardboard forms of 8-10 mm wall thickness have multiple overlapping paper layers and are protected from moisture penetration on the inside and outside. They are, however, very vulnerable to moisture penetration at the ends. Plastic and galvanised sheet steel forms are spirally wound from one layer of material. Neither of these two have any problems with moisture.

All three types impart a spiral pattern to the concrete face. The pattern from the cardboard tube form is the least pronounced of the three, but is still evident from quite a distance. If this column is to be sandblasted for a depth of at least 2 mm, this pattern usually disappears. With sheet metal and plastic forms the spiral pattern is usually too deep for removal by sandblasting.

They all require support to maintain plumb and vertical line. The metal and plastic forms also require support to maintain their circular shape. As Figure 5.45 indicates this may involve bracing to intermediate levels as well as the top.

Reference should be made to the manufacturer's technical literature to determine the maximum spacing of the braces. The determining factors are the column diameter and overall height.

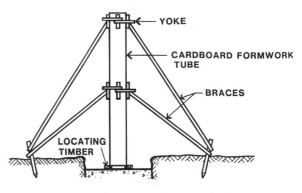

Figure 5.45 - BRACING TO A CIRCULAR COLUMN FORM

Attention must be paid to the way in which the bracing is attached to the column form. Figure 5.46 shows the simplest type of yoke consisting of a square frame and four wedges. With care this can work with the cardboard forms, which are quite stiff and not readily distorted by the wedges.

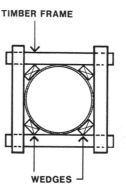

Figure 5.46 - YOKE FOR A CARDBOARD TUBE FORM

For the more flexible plastic and sheet metal forms, this type of yoke is unsuitable as these thin forms are readily distorted out of round when the wedges are tightened. Figure 5.47 shows the sectional plan of a yoke that is suitable for flexible forms.

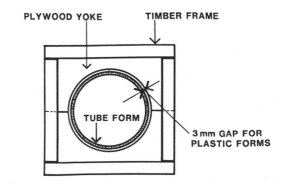

Figure 5.47 - YOKE FOR PLASTIC AND METAL FORMS

Two pieces of plywood are cut to halves of the round form shape and cleated together. For the metal forms they can fit tightly to the form. When used with plastic forms a clearance gap of approximately 3 mm should be provided to allow for the stretching of the plastic form caused by the fluid concrete pressure.

The direction of erecting the plastic spiral forms is marked on the form. The form is erected with this arrow pointing to the top. This ensures that, as the level of the fluid concrete rises, its pressure aids the sealing of the tongue and groove junctions in the plastic spiral.

The precaution to be observed with cardboard forms relates to the water absorption which was noted previously. If there is water lying on the top of the concrete at the base of the form it will be readily absorbed up into the cardboard. This reduces the tensile strength of cardboard to almost nothing. Form failure during the concrete pour will inevitably occur. The tops of these concrete bases must be dry, and if the forms are to be left overnight, they should be lifted above the maximum possible water level. The cardboard forms must be inspected before the pour.

Where these types of forms are to be used, care should be taken with the reinforcement details. It is normal for the reinforcement to be tied in place and then the tube form hoisted over the top. If the tops of the vertical reinforcement are bent over to lap into a floor structure, the form cannot be installed in this way; it must be installed simultaneously with the reinforcement. This is difficult and there is the strong possibility of damage to the formface.

The stripping of both the metal and the plastic forms is simply a matter of unwinding the ribbon of material from the top. For the multi-layer cardboard forms unwinding is difficult. They are usually stripped off by making at least one vertical cut through the cardboard and a shallow matching cut on the opposite side of the form. The cutting and removal procedure is illustrated in plan section in Figure 5.48

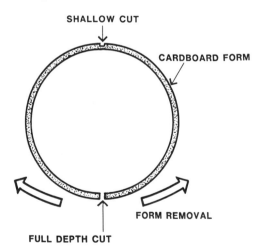

Figure 5.48 - STRIPPING A CARDBOARD COLUMN FORM

The cardboard form can then be cracked and peeled off in two pieces. These three types of forms can only be used once and after this nothing of value can be recovered.

One type of proprietary round column form is both re-useable and adjustable in diameter. In addition, it has the advantage of being suitable for the production of high quality concrete surfaces. The form consists of a large sheet of tough flexible plastic that is rolled into a multi-layer cylindrical shape. A plan section is shown diagramatically in Figure 5.49.

PLASTIC SHEET ROLLED TO CYLINDRICAL SHAPE

Figure 5.49 - COLUMN FORM OF ROLLED PLASTIC SHEET

The size of the column form can be selected, within a range, by the formworker. The tightly rolled sheet is then held at that diameter, and against internal fluid concrete pressure, by a strap and buckle system. These are similar to those shown in Figure 5.27. Reference should be made to the manufacterer's literature for data to determine the spacing of the straps relative to column diameter and the height of the pour.

Similar to the spirally wound columns, the rolled sheet requires yokes at bracing points to maintain its shape. For slender column forms, bracing is also needed at intermediate heights.

Stripping is very simple. The straps are released and the sheet unwound. Although its first cost is high this forming method can often be the most economical due to its versatility and the large number of re-uses that are possible.

Convential column forming techniques for plywood and timber can also be adapted to round columns. Figure 5.50 shows the sectional plan of one example of this type of forming.

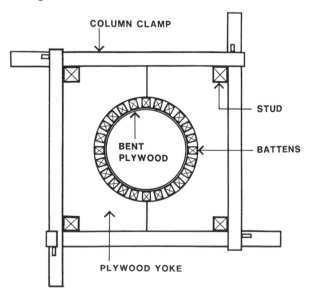

Figure 5.50 - TIMBER AND PLYWOOD COLUMN FORM

Thin plywood is curved to shape as two semi-circular pieces supported on closely spaced vertical battens. As noted in the Table in Chapter 4 on curved wall forms, plywood as thin as 4 mm can be used for diameters down to 355 mm. The load from the battens is carried onto thick horizontal plywood yokes which have square timber studs at each corner.

Proprietary column clamps provide the needed tensile capacity around the outside of the yokes to resist fluid concrete pressures. The size of the plywood yokes must be selected, relative to their thickness and strength grade, to provide the needed bending strength between the corners.

Round column forms can also be fabricated in two halves from steel or glass reinforced plastic (GRP). A typical section for a steel form is given in Figure 5.51. GRP forms are similar but the wall, flange and stiffeners thicknesses are greater. In both cases the fabricated sections would incorporate fixings for braces. These are usually fitted to the stiffening ribs.

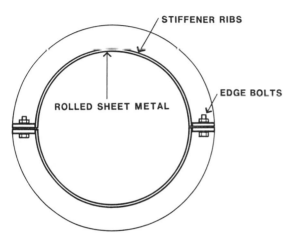

Figure 5.51 - CIRCULAR STEEL COLUMN FORM

The connection of the two halves is either by bolts or proprietary wedges. As noted earlier in this chapter, wedges enable faster assembly and stripping of the form.

Advantage can be taken of the natural springiness of high tensile steel to fabricate a one piece round column form. Figure 5.52 shows this type of column form before installation. The high tensile steel, usually 450 Grade steel sheet, 2mm or more thick, is rolled to a diameter 12 mm greater than the column size and the gap at the joint fabricated to 38 mm. Note that the angles at the joint are positioned to mate properly to each other when the gap is closed.

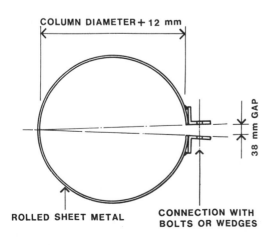

Figure 5.52 - SINGLE PIECE STEEL COLUMN FORM

The form is installed by hoisting over the reinforcement and the joint clamped tightly closed with bolts or proprietary wedges. Because sheet metal is flexible, the attachment for the braces should incorporate a yoke that maintains the circular shape.

Stripping starts with the removal of the bolts or wedges. With this release, the form springs open to a size larger than the column, and can be withdrawn vertically off the column. Note that some surface abrasion may occur at this time.

GENERAL DETAILS

Column Kickers and Kickerless Construction

Similar to kickers for walls, column kickers serve to inhibit the loss of fines at the base of the form and position the column form in its correct place.

The accuracy of the column position may also be affected by the accuracy of the positions of the column starter bars and the permitted variations in the concrete cover to the reinforcement. It also depends on the method of lapping the bars.

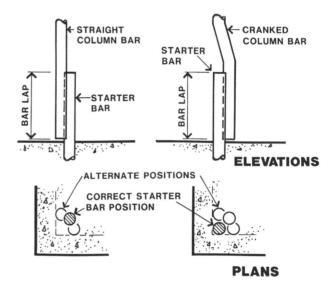

Figure 5.53 - ACCEPTABLE STARTER BAR POSITIONS

Figure 5.53 shows the permitted variation in acceptable starter bar positions for both straight and cranked laps. If the starter bars are outside these acceptable positions, there is the strong possibility that the column will also have to be incorrectly placed.

For the straight laps, the permitted positions are all on an arc which is located around the straight column bar. With the cranked laps, the correct position of the starter bar is directly below the column bar. However, the starters may also be placed on an arc around this position. Clearly, cranked laps give the greater tolerance in starter bar position.

For columns located in the middle of floor slabs or large footings it is difficult to accurately position the form for the kicker so that it can be poured integrally with the slab. If a kicker is wanted it is possible for it to be formed and poured after the slab pour. Figure 5.54 shows the section of such a kicker and its formwork.

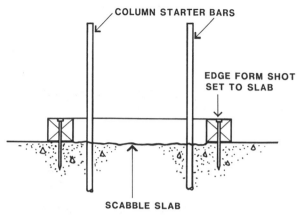

Figure 5.54 - KICKER FORMED AFTER SLAB POUR - 1

The kicker edge forms are timbers shot-set to the slab. The slab is scabbled to aid the bonding of the new concrete and a light mesh, usually having 4 mm rods at 100 x 100 mm spacing, is placed in the concrete of the kicker.

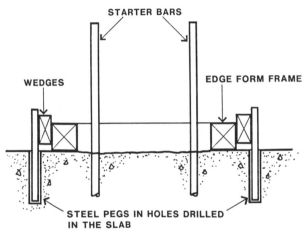

Figure 5.55 - KICKER FORMED AFTER SLAB POUR - 2

Figure 5.55 illustrates the kicker edge forms wedged between loose steel pegs placed in holes drilled in the slab. Alternatively, as shown in Figure 5.56, the column starter bars can be used to wedge the kicker edge form frame in place.

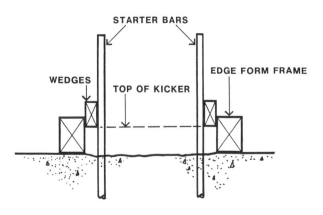

Figure 5.56 - KICKER FORMED AFTER SLAB POUR - 3

Similar to walls, column forms can be located without concrete kickers. Generally, the techniques used for

columns are derived from wall forming techniques. The simplest one, as shown in the section through the base of the column form in Figure 5.57, involves shot-setting locating battens onto the top of the footing or floor slab.

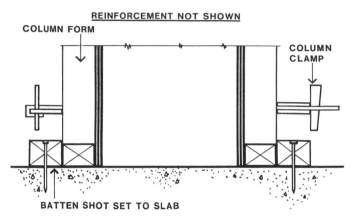

Figure 5.57 - KICKERLESS COLUMN CONSTRUCTION - 1

These battens can hold the column form both in position and with its faces true to line, that is, the formwork cannot twist. Internal locating systems can control the column position, but are not as effective in eliminating twisting. The first example of this latter type is illustrated in Figure 5.58. It is similar to the device for wall forms shown in Figure 4.69.

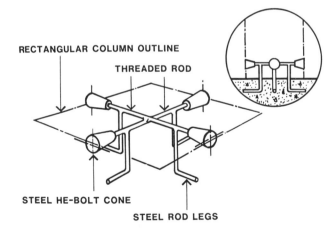

Figure 5.58 - KICKERLESS COLUMN CONSTRUCTION - 2

Two threaded rods are fitted with steel he-bolt cones at their ends and welded to steel rod legs. The device is cast into the slab. Screwing the cones to position gives the final adjustment to the line of the column faces. The hexagonal recesses in the cones should be filled or taped over to avoid grout penetration. After stripping the forms, the cones can be screwed out.

This method is often considered to be inconvenient as the device obstructs the finishing of the surface of the floor slab. A method that is installed after the slab pour is given in Figure 5.59.

Small eccentric circular fibre cement disks (usually 40 - 50 mm diameter) are screwed to the concrete surface. By rotating the disks their outer edges can be accurately set to the lines of the column. There are two disadvantages: there is no control over twisting of the form and the disks result in an effective reduction of column concrete area.

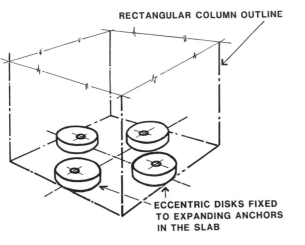

Figure 5.59 - KICKERLESS COLUMN CONSTRUCTION - 3

Kickerless forming methods can result in grout loss between the bottom of the form and the uneven concrete surface. As columns are load bearing concrete members this is usually unacceptable. Figure 5.60, which is a refinement of Figure 5.57, shows one method of sealing the gap, between the top of the slab and formwork, to prevent this loss.

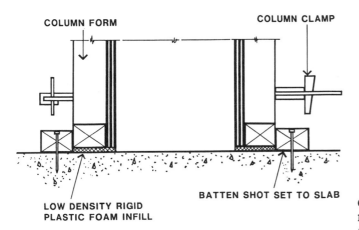

Figure 5.60 - SEALING THE BASE OF THE FORM

Low density rigid plastic foam is bonded to either the underside of the form or the top of the concrete. The bonding is needed to prevent the foam being dislodged during the erection of the form. Upon erection, the weight of the form squashes the foam. With care this is method that can effectively fill small gaps.

Marking the Top of the Pour

It is the formworker's responsibility to mark the level of the top of the concrete pour. The method used must be able to survive the pouring of concrete, be clearly visible after being splashed with grout, and show the tolerance limits of the pour height.

In the case of a concrete column supporting steel beams, the column height tolerance will be controlled by permitted variations in base plate grout thickness. Figure 5.61 shows the elevation of a simple steel beam to column connection.

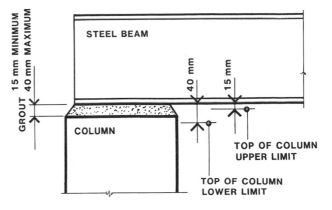

Figure 5.61 - STEEL BEAM SEATED ON A COLUMN

For a column supporting a slab or a concrete beam it is usually permissable for the column to extend part way into the concrete cover of the bottom reinforcement. Also, at its lowest level, it must at least penetrate into the plywood of the slab soffit form. Figure 5.62 shows an example of this situation.

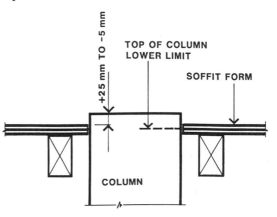

Figure 5.62 - TOLERANCE COLUMN TO SLAB

Here the tolerance range is 30 mm. The simplest way of marking this on the column form is to drive pairs of nails into all four internal faces of the column form. (Figure 5.63) The concretor then fills the column up to or above the lower nail but always ensuring that the upper nail remains visible.

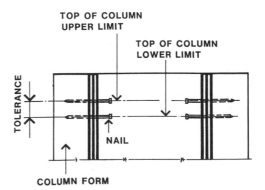

Figure 5.63 - MARKING THE COLUMN TOP - 1

This method does not produce a neat line at the column top. If one is needed then the method shown in

Figure 5.64 is suitable. The tolerance indicated is of the same order as that for the previous example.

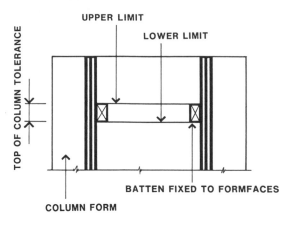

Figure 5.64 - MARKING THE COLUMN TOP - 2

A batten of height equal to the tolerance range is fixed to all faces of the column form. The concrete pour must at least reach the bottom of the batten but not extend beyond its top.

Column Formwork Bracing

Column form bracing performs two functions. Firstly, it must maintain the accuracy of the column form position and plumb so that it is within tolerance. Its second function relates to the possible results of forces acting on either the column formwork or the bracing. The forces may be wind or impact. These impact forces can occur from the collision of concrete skips or crane hoisted bundles of material. They can act in any direction.

It is not expected that the form and its bracing can always withstand these impacts without damage or misalignment. What is important is that there must not be catastrophic collapse or the generation of falling debris.

The descriptions and illustrations given below on column bracing generally relate to tall and slender columns. For very large and heavily reinforced columns, such as those in the lower levels of high-rise buildings, the widths of the column form and the robustness of the reinforcement may be such that only accuracy adjustment devices at the form base are required. In these cases, reinforcement and form assembly may be sufficiently strong within itself to resist any expected impact.

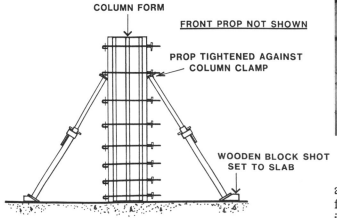

Figure 5.65 - UNFIXED COLUMN FORMWORK BRACING

Column bracing which is not attached to both the form and the base cannot comply with the requirement not to generate debris under impact forces. Figure 5.65 shows this situation. Only the propping in one of the two required directions is shown.

This method of column bracing is commonly done, but it is poor practice. The telescopic props are simply wedged between the column form clamps and the blocks, shot-set to the floor slab. Any impact on these props, even a small one, could dislodge the prop. In falling it could cause injury; if the column form is at the edge of the floor slab then danger to the public occurs.

However, telescopic props can be suitable for column bracing if they are fitted with connection devices like those shown in Figure 5.66.

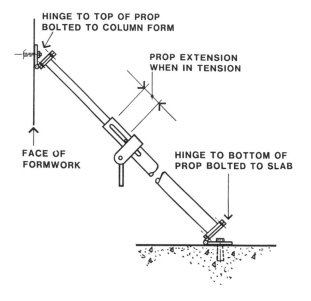

Figure 5.66 - FORM BRACING FIXED AT BOTH ENDS

The bolted hinges at the ends of the props help resist their dislogement under impact. If a tension force acts on the prop, the lock pin will slide to the other end of the slot, and the prop then has resistance to tension loads. During this movement, a good deal of the energy of the impact will have been absorbed.

Steel Column Formwork with Bracing Attached.

If drilling of the slab for a fixing is not permitted then a complete bracing frame can be bolted to the column form. These can be purpose built fabrications or incorporate proprietary adjustable struts as shown in Figure 5.67. A purpose built bracing assembly would require an adjustment device either in the sloping brace or acting vertically at the bearing point on the slab.

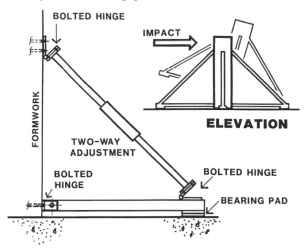

Figure 5.67 - BRACING FRAME BOLTED TO THE FORM

Impact on this assembly may cause horizontal sliding which will be resisted by the column reinforcement. If the horizontal impact is at the top of the form, the whole assembly will tend to rotate about the outer base of the braces and the column form will tend to lift upwards. This will be resisted, in part, by the weight of the assembly and the forces needed to bend and distort the reinforcement. Note that overtightening of the sloping brace adjustments will lift the column formwork off its base and severe grout loss will result.

For internal columns there must be enough braces to resist forces in any directions. For four sided (rectangular) columns, four sets is the convenient number due to fixing considerations. For round columns three equally spaced braces are usually adequate.

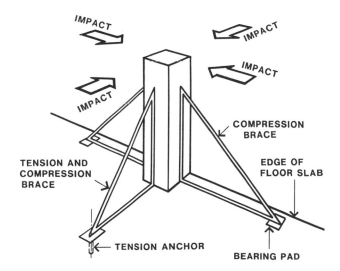

Figure 5.68 - BRACING FOR EDGE OF SLAB COLUMNS

It is not usually possible to provide bracing connected to all sides of a column located at the edge of a suspended floor. Figure 5.68 shows one solution to the problem. To resist forces acting both inwards and outwards, a tension-compression brace is fitted to the inner side and bolted to the slab.

If bolting is not permitted then a large counter-weight can be attached to the assembly. This is less satisfactory than the bolting to the slab as the counterweight becomes part of the group of items that are potential debris.

Integral Pour of Columns with Slab

All concrete suffers from shrinkage and this is greatest in the first days after the pour. If it is proposed to pour the columns at the same time as the slab, then this shrinkage will tend to shorten the columns and compress the column formwork. Restraint of the column shrinkage will cause tension in the concrete and may lead to cracking.

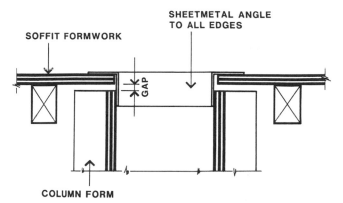

Figure 5.69 - FORMWORK JOINT FOR INTEGRAL POUR

To avoid the column formwork being jambed in place, difficult to strip and possibly damaged, a movement joint should be provided. Figure 5.69 shows an example of a movement joint between the column and slab forms.

Sheetmetal angles are fitted to the four edges of the penetration in the soffit form. The vertical legs of angles bridge the gap (usually 10 mm) to the column form. When shrinkage induced movement occurs the angle usually slides down inside the column form. If the interface is too tight for sliding, the thin sheetmetal will readily buckle and vertical movement will not be prevented.

Hoisting Column Forms.

For the individual formfaces of conventional column formwork, the hoisting point can simply be a bolt between two studs near the top of the form. (Figure 5.70)

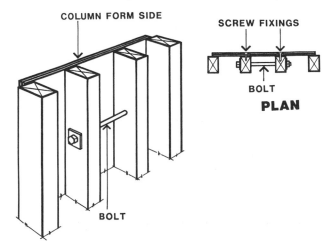

Figure 5.70 - SIMPLE FORMFACE HOISTING POINT

Nail fixing of the plywood to these studs will not be adequate to cater for the impacted loads of crane hoisting. At the least, screw fixing will be needed. The bolt between the two studs should be adequate to cater for the weight of the formwork. At the least it should be 12 mm diameter and locknutted at the end. The practice of using a plain rod for the bolt must be avoided. With a high impact load it can pull out, causing an accident.

If the individual formface is fitted with a bolted bracing assembly this can be used for slinging. Figure 5.71 illustrates the sling placed under the top end of the raking brace.

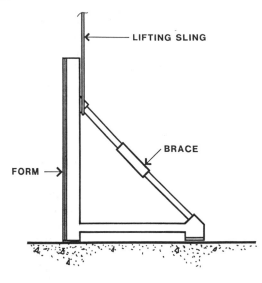

Figure 5.71 - HOISTING FROM THE BRACING FRAME

Figure 5.72 shows a method of providing a hoisting fitting to a horizontal waler column formface. The lifting eyebolt should be fitted with an adequate bearing washer at its lower end. The walers must be bolted to the studs. The bolt heads with washers are shown recessed into the studs to avoid penetration of the formface.

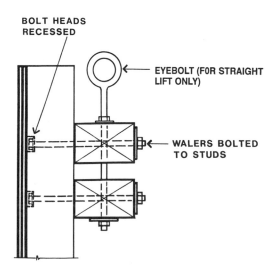

Figure 5.72 - COLUMN FORMWORK LIFTING POINT

These eyebolts can also be used for the hoisting of the completely assembled column formwork unit. To avoid conflict with the vertical column reinforcement, a spreader should be fitted between the lifting cables. (Figure 5.73)

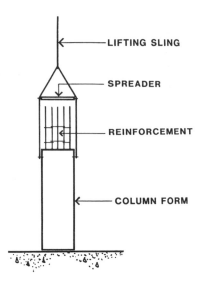

Figure 5.73 - CRANE HOISTING THE COLUMN FORM

Hinged forms can also be lifted by hoisting points on the top walers. This is illustrated for a single hinge form in Figure 5.74. Hoisting for installation over the reinforcement can take place with the form clamped closed.

Alternatively, it can be wedged partly open when being retracted in stripping. This is done to avoid abrasion between the faces of the concrete columns and the formfaces.

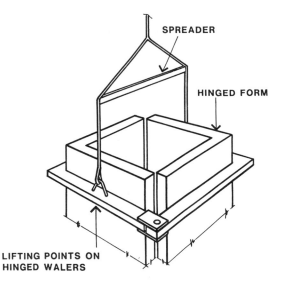

Figure 5.74 - HOISTING A HINGED COLUMN FORM

More efficient hoisting of complete forms can be achieved with a purpose made steel lifting frame. As shown in Figure 5.75 the frame must be dimensioned to avoid any conflict with the extended column reinforcement.

The security of the formwork assembly during hoisting is paramount. One necessary precaution is to provide the lifting frame with steel locating pegs that penetrate the walers. These pegs must be sufficiently robust and long that a disturbance or impact to the form during hoisting does not dislodge it.

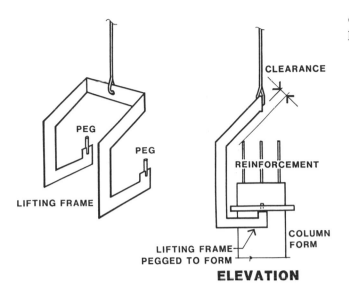

Figure 5.75 - COLUMN FORMWORK HOISTING FRAME

Access Platforms

Formworkers require access to the upper parts of the column form at all stages of the construction process. During formwork erection access is needed for the installation of ties and clamps, during Stage 2 access at the top is needed for concrete placement and vibration and finally, for the safe removal of clamps and ties and the connection for hoisting, access is needed during stripping.

Scaffold Frame for Access to Column Formwork

At the simplest an access platform can merely be a standard scaffolding frame planked out at the appropriate level, fitted with a handrail and provided with an access ladder. If it surrounds the column it will give access to all sides of the formwork. The principle problem with this type of platform is that moving it to another location usually involves dismantling the frame.

If there is sufficient repetition of use, a purpose-made access platform can both the most efficient and economical. Ideally, it will give access to all sides of the form, be hoistable in one piece for ease and speed of relocation, be fitted with safety handrails and an access ladder and be proportioned to not to tend to tip over under impact forces. Ease of relocation implies that it does not interfere with the column bracing. Figure 5.76 shows the diagramatic arrangement of a typical purpose made platform that meets these criteria.

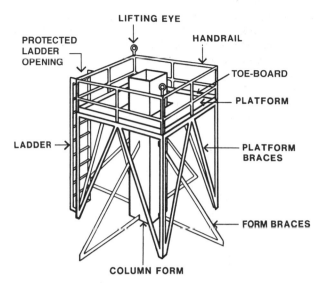

Figure 5.76 - PURPOSE MADE ACCESS PLATFORM

This access frame serves all stages of construction and can be speedily moved from column to column. On large projects it is sometimes economic to have a special platform exclusively for Stage 2, the concrete placing. Figure 5.77 shows such a platform; in this example it is carried on the formwork. For these cases it is normal to have a separate access ladder. Concrete placing does not require access to all sides of the form and here one side is used for a concrete pouring chute built into the platform.

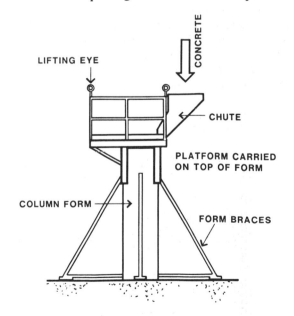

Figure 5.77 - PLATFORM CARRIED ON THE FORMWORK

A platform of this type can have a smaller floor area than that of the all-purpose unit. It will also be lighter and more easily relocated during the progress of the column pours. It use, however, is usually restricted to columns of the same size because it fits onto the formwork. On the other hand, the all-purpose platform normally has an oversize deck opening to suit a range of column sizes.

Columns at the edge of a suspended floor present a work hazard. The provision of access all round the formwork calls for a cantilever platform. Figure 5.78 diagramatically shows a platform with a stabilizing safety counterweight. The size of the counterweight must take into account the structural design case of the workmen all being on the outer edge of the platform, with the simultaneous action of wind and impact.

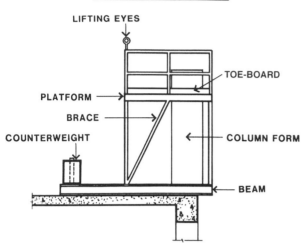

Figure 5.78 - EDGE COLUMN ACCESS PLATFORM

It is essential that this counterweight be securely fixed. Its dislodgement could easily lead to the platform toppling over the edge if the formworkers were on the outer side of the working area. A strong pegging arrangement, as shown in Figure 5.79, can provide a secure fixing for the counterweight. Impact cannot dislodge this counterweight and it is made too heavy to be inadvertently removed by workmen.

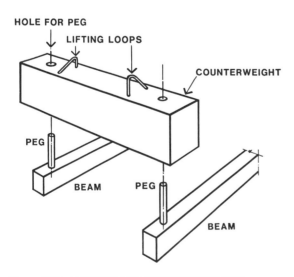

Figure 5.79 - PEGGING OF COUNTERWEIGHT

Following the principle of stability at all times, the platform must be proportioned to be stable without the counterweight. The platform will be hoisted into place without the counterweight. Because these counterweights are heavy, it is common for them to be hoisted into position by crane after the platform is in place.

Gang Forming

Where a regular pattern of columns or columns and beams occurs, economies are often possible by forming them together in groups. This is known as gang forming and is simply a number of individual formwork units fixed to the same support frame. This is common practice for building facades of high-rise buildings. Figure 5.80 shows an example of a case suitable for the gang forming of beams and columns together. However, in this chapter information will be confined to the gang forming of columns only.

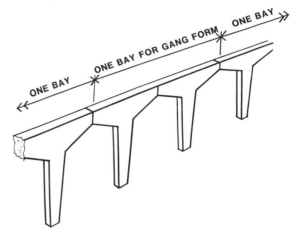

Figure 5.80 - SUITABLE CASE FOR GANG FORMING

By forming columns in groups, the time and cost of formwork erection and stripping can be reduced. To maximise this advantage the design of the framing of the gangform must take account of the inaccuracies that can occur with multiple use.

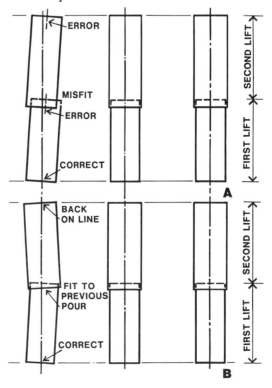

Figure 5.81 - EXAMPLE OF USE OF COLUMN GANG FORM

This topic of progressive errors was illustrated for the simple case of horizontal grooves on wall forms in Figures 4.92 and 4.93. For gang formed columns the problem of progressive errors is more complex.

Even if the most stringent control on accuracy of the gangform is maintained during its fabrication, small dimensional deviations will occur. Similar to the case of repetitive use of wall forms, misfits can occur due to progressive errors from the repetitive use gang forms. Figure 5.81, A and B, gives a simple illustration of the use of a three column gangform .

The gangform is shown in Figure 5.81-A with the left hand column form slightly out of plumb. When the gangform is hoisted for the second lift, a misfit occurs between the bottom of the column form and the top of the previously poured concrete column. The gangform needs a means of adjustment so that each column form can be adjusted to fit to the top of the previous pour and also be adjusted back to the correct position at the top of the column form. (Figure 5.81-B)

Errors can occur on both axes at the top of the columns and these require a means of adjustment. At the bottom of the columns, two axes adjustments are required to enable mating of the base of the individual forms to the top of the previously poured columns. Figure 5.82 shows the directions of these adjustments.

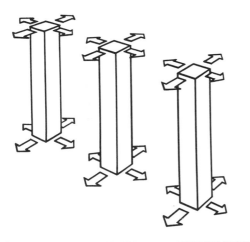

Figure 5.82 - GANG FORM ADJUSTMENTS NEEDED

Further, the column forms may not be the correct distance apart, top and bottom. The correction of this will use the same adjustments.

Within each column form, errors of angular alignment in relation to the total column group can occur. Figure 5.83 indicates possible column twist which must be adjusted at both the top and bottom of the form. The twist is shown in the left hand column.

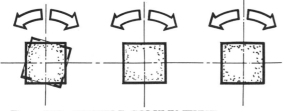

Figure 5.83 - POSSIBLE COLUMN TWIST

Finally, the column boxes may be out of square. (Figure 5.84-B) If the forms are fixed on one common

face to their frame, simple turnbuckle adjustments can be used to correct out-of-square forms (Figure 5.84-A).

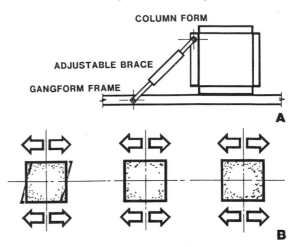

Figure 5.84 - COLUMNS OUT OF SQUARE

The need for all these adjustments does not mean that gangforms must be prohibitively complex and expensive. Even in extreme cases the adjustments rarely exceed 4 mm. If it does exceed 4 mm, it usually means that the original gangform fabrication was very inaccurate or insufficiently rigid.

The expected range and direction of the adjustments, between the individual column forms and the gang frame, can usually be catered for with the simple adjustable bolted joint shown in Figure 5.85. The Figure shows a part section on the bolt centre-line.

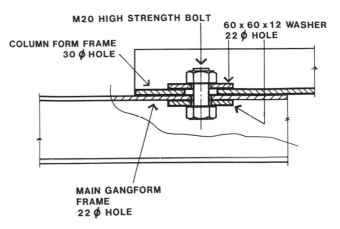

Figure 5.85 - FRICTION GRIP ADJUSTMENT JOINT

This is a friction grip joint. The oversize holes in the frames to each column form permit 5 mm adjustment in any direction. The holes in the gangform frame are a neat fit to the bolts. To adjust, the bolt is slightly slackened and the column form frame can be tapped across to its correct position. The bolt is then tightened. The bolts should be high strength to avoid fatigue failure from repeated tightening.

Figure 5.86 gives a diagramatic plan view of a column form that incorporates turnbuckles for out-of-square forms, and this bolted friction joint for all other ajustments between the individual column forms and the gangform frame.

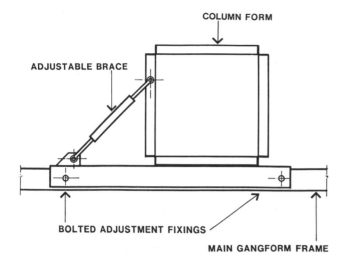

Figure 5.86 - PART PLAN OF COLUMN GANG FORM

The diagramatic view of Figure 5.87 shows this type of construction incorporated in the whole gangform. As noted, there is a need for lateral braces to control the plumb and twist of the whole assembly. If the longitudinal beams of the gangform frame are sufficiently rigid laterally, only two double acting adjustable braces are needed, one at each end of the gangform.

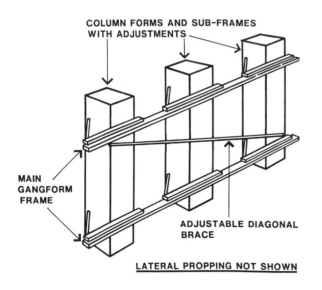

Figure 5.87 - COLUMN GANG FORM ASSEMBLY

Although not shown in any of the illustrations, it is normal practice for these gangform assemblies to be fabricated with access platforms and ladders permanently attached.

An important factor in the cost of use of column gangforms is the ease and speed of their stripping and subsequent re-erection. The adopted cycle of use of the gangform has a controlling effect on this. The sequence of this cycle is often a predetermined function of the detailed design of the gang form. A comparison of three different cycles for external columns can be made.

In Figure 5.88 the gangform is lifted into place after the completion of the floor pour. After the column pour, the form is stripped in one or several parts, and lifted vertically off to a separate place of storage.

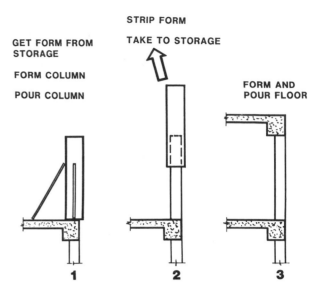

Figure 5.88 - SEQUENCE OF GANG FORM USE - 1

After the floor pour, the cycle repeats. A constraint on the use of this system is the need for a substantial formwork storage area. Most high-rise buiding sites do not have this facility.

If the gangform is designed so that it does not need to be stripped right after its pour, then the cycle shown in Figure 5.89 can be used.

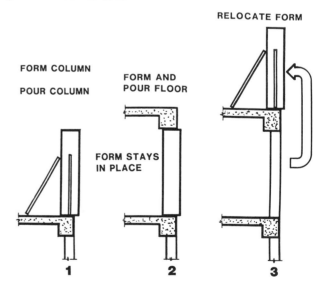

Figure 5.89 - SEQUENCE OF GANG FORM USE - 2

The cycle starts with the forming and pouring of the columns. The details at the top of these column forms are arranged to permit the forming and pouring of the slab without the columns having to be stripped. After the floor pour the column form is stripped open. It is then moved outwards and taken outside the buiding up to its next position for re-erection for the next cycle.

This is the type of cycle that is often used when hinged forms are mounted on the inner face of self-climbing perimeter safety screens. Here, the column forms are securely fixed to the screen so they can be opened for inspection of the concrete before the floor is poured.

A variation of this cycle is shown in Figure 5.90.

RELOCATE FORM

FORM SLAB AND
POUR WITH COLUMN

POUR COLUMN
WITH SLAB

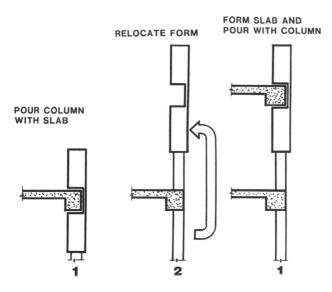

1   **2**   **1**

Figure 5.90 - SEQUENCE OF GANG FORM USE - 3

In this case the column construction joint is located midway between floors, and the column pour is simultaneous with the floor pour. After the pour, the column gang form is stripped and re-erected on the top of the previously poured columns. The floor is then formed for the next pour.

Like the previous case this system is stripped to the outside and the type of gangform used must enable this to be done.

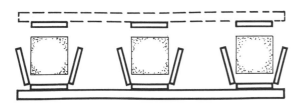

Figure 5.91 - GANG FORM STRIPPING SYSTEM - 1

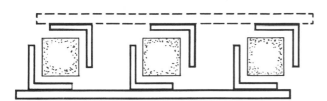

Figure 5.92 - GANG FORM STRIPPING SYSTEM - 2

Figures 5.91 and 5.92 show two examples of diagramatic plans of two-part gang form systems. Each example is effectively two gang forms. Part of the forms are on one gang frame, the remainder are on another.

These would only be suitable for the cycle shown in Figure 5.88, where the columns are stripped immediately after pouring, and prior to the commencement of the floor formwork. The assembly which is stripped inwards is large and needs to be lifted out before the slab is formed.

For cycles that have a need to strip the whole gangform assembly outwards, the arrangements shown in Figures 5.93 and 5.94 are suitable.

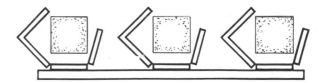

Figure 5.93 - GANG FORM STRIPPING SYSTEM - 3

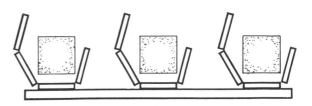

Figure 5.94 - GANG FORM STRIPPING SYSTEM - 4

The system shown in Figure 5.94 is the most commonly used one for mounting on the inner face of a self-climbing perimeter safety screen. Having three hinges, each form can be folded back flat against the screen so that the screen can climb past the edge of the floor.

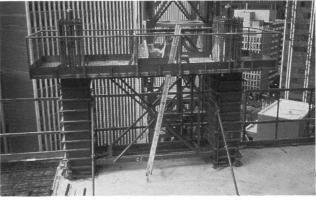

Two Column Gang Form with Access Platforms

SPECIAL COLUMN SHAPES

All the information in this chapter so far, has related to either rectangular or circular columns. However, in a minority of cases, special shapes are sometimes called for. By the careful adaption of conventional forming methods the costs of these special shapes can usually be controlled.

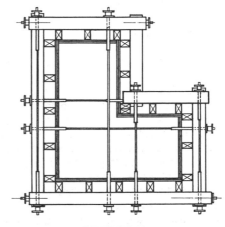

Figure 5.95 - FORMS FOR 'L' SHAPED COLUMN

A simple example of this is an 'L' shaped column. As Figure 5.95 shows, the use of conventional techniques exactly following the column perimeter leads to a very complex form. To cater for the fluid concrete pressures on the 'L' shape, an unusually large number of internal ties are needed. The waler arrangements are similarly complex.

One method, that eliminates much of this complexity, involves the use of a rectangular form that fits the outer lines of the 'L' shape. The re-entrant corner is formed with a rectangular packer form that fits inside the outer form. This is shown in Figure 5.96. This inner packer form is subject to the same fluid concrete pressures as the outer rectangular form.

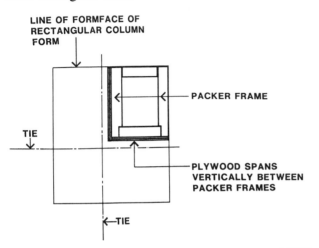

Figure 5.96 - PACKER FORM FOR 'L' SHAPED COLUMN

Packer forms are a technique suitable for a wide range of shapes. Figure 5.97 shows two examples.

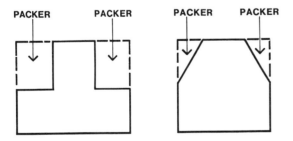

Figure 5.97 - OTHER SHAPES FOR PACKER FORMS

The rectangular form can also be used to contain rounded shapes. Figure 5.98 shows circular proprietary plastic spiral-wound column forms squashed into a rectangular form box to create a column with straight sides and rounded corners.

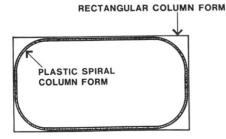

Figure 5.98 - PLASTIC FORM IN A RECTANGULAR SHAPE

This method, by careful control of the diameter of the circular plastic form, can also give semi-circular ends. Another method that achieves semi-circular ends is shown in Figure 5.97.

This uses halves of cardboard spiral-wound proprietary circular forms at each end, with plywood packers, of the same thickness, between. There are two essential precautions. Firstly, the cut edges of the cardboard must be sealed against moisture. Secondly, the space between the cardboard and the corners of the rectangular formwork box must be solidly packed.

One method successfully used is to pack the space with a weak, relatively dry, mortar mix or no-fines concrete.

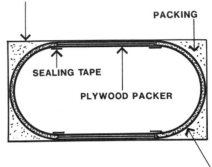

Figure 5.99 - COLUMN WITH SEMI-CIRCULAR ENDS

In some cases a purpose made section of the forms can be used in conjunction with proprietary products. Figure 5.100 shows the sectional plan of the forms for a column with two straight sides at right angles and a curved one joining their ends.

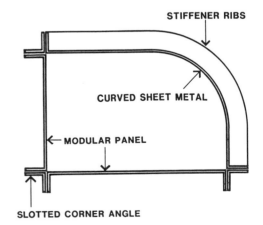

Figure 5.100 - COLUMN WITH ROUNDED CORNER

Column Form with Rounded Corner and Straight Sides

The straight sides are formed from modular panels with proprietary slotted angles and connection wedges at all corners. The special panel has a rolled sheetmetal face with slotted end plates and closely spaced stiffeners between the ends. These stiffeners carry the bending load from the fluid concrete pressure, and tensile and bending loads from the ends of the flat modular panels.

CHAPTER 6:
SOFFIT FORMS

Soffit forms can be defined as level, or near level, single faced formwork surfaces used to be the mould for the underside of reinforced concrete structures such as suspended floor slabs. These form surfaces are walked on by the site personnel, are used as a storage area for formwork components and reinforcement, and provide a working area for the formworkers and other following trades. Its function, at all construction stages, as a work area as well as a mould for the fluid concrete makes safety a paramount consideration. The completed form should have guardrails at its perimeter and be provided with adequate access ladders or stairs.

Soffit forms are subject to the construction loads of workers, materials and equipment, possible impact from hoisted loads and concrete buckets, and the environmental loads of wind and water.

These considerations were discussed in the Overview of Chapter 1 and related to the three construction stages and the associated design requirements of **STABILITY**, **STRENGTH** and **SERVICEABILITY**. As illustrated in Figure 1.22, sometimes more than one stage of construction can occur at the same time. As a preamble to information on the techniques of soffit form construction, the relevant aspects of the three construction stages will be restated here.

Stage 1 is the construction of the form, and making it ready for the placing of concrete. At this time it is light in weight, and very susceptible to the destabilising effects of wind, horizontal impact and incorrectly placed loads. Where the works are located in high or very exposed locations (Figure 1.26) the wind forces can be very great.

The absence of effective bracing (Figure 1.08) at any time can lead to sidesway. Sliding of the whole form structure (Figure 1.03) may result if the lateral restraint at the footings is insufficient. A tall and light form assembly may tend to overturn (Figure 1.05) and guy ropes may be needed (Figure 1.06).

Stability problems can occur within the form structure. Unequal loading can lead to framing members deflecting upwards off their supports (Figure 1.07). All of these situations are the seeds of failure.

Materials Stacked on a Soffit Form

The vertical loading, which the form must have adequate strength and stability to cater for, consists of the formworkers, their tools, and the materials that they are going to use to complete the work ready for concrete placement. It is common practice for a part of the total soffit form to be completed and then the materials for the remainder of the soffit; joists, bearers and plywood, stacked on this completed part. The completion of the form progresses out from there.

Reinforcement Stacked on the Formwork

When a significant portion of the form is completed and clear of stacked materials, other following trades can start. These will usually be reinforcement tying, and installation of electrical conduits and built-in components for other services. For reinforcement the bundles may weigh 5 tonnes, or more, and these are often crane hoisted onto the form. Figure 1.20 showed a case where the progressive construction of the form, of alternating rows of frames and props, can be a potential hazard. If the hoisting of a reinforcement bundle precedes the erection of the intermediate rows of props, and the bundle is placed over this longer joist span then the form will fail.

Even if the form is correctly framed for the high material loads, failure can occur from vertical impact caused by the rapid lowering of material bundles.

Serviceability relates to the effective control of deformation, deflection and movement, under the expected loads. In general, this is not a Stage 1 problem, except that excessive movement, even if transient, must not result in permanent misalignment or affect the tightness of the joints of the formwork.

Of all the three stages, it is **Stage 2**, when the concrete is being placed, that results in the highest loading and poses the greatest danger of structural failure. The horizontal forces that affect stability are wind, acting on both the supports and the edge of the formwork, horizontal impact from hoisted concrete buckets (Figure 1.23/2) and the effects of construction activity, e.g. surge in pump lines.

The greatest vertical load is usually the concrete and its effects are increased by mounding, either inadvertent

(Figure 1.18) or deliberate (Figure 1.19). To this can be added the vertical impact from concrete buckets striking the form (Figure 1.23/3) and the weight of workmen and their equipment.

Stage 3 is defined as the period after the pour when the form structure remains in place until the concrete has attained enough strength to support itself and any loads placed on the freshly poured slabs. These extra loads can be formwork components for the next level or materials stored on multi-storey slabs for other trades to use after the formwork is all completed (Figure 1.21).

The much enhanced load effect from this requires detailed and careful consideration by the design engineer for the building. Materials must not be placed in this way without permission.

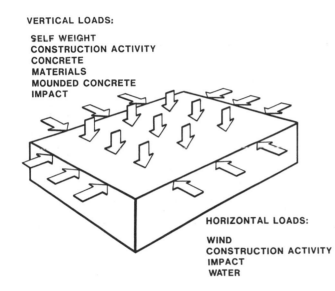

VERTICAL LOADS:

SELF WEIGHT
CONSTRUCTION ACTIVITY
CONCRETE
MATERIALS
MOUNDED CONCRETE
IMPACT

HORIZONTAL LOADS:

WIND
CONSTRUCTION ACTIVITY
IMPACT
WATER

Figure 6.01 - SOFFIT FORMWORK LOADING

Figure 6.01 diagramatically gives the overall scene of the loads that must be considered in the design and construction of horizontal soffit forms. These loads may all act simultaneously, or individually. In some cases they may act on the whole form area, or just on part of it. Horizontal loads can act from any direction. The formwork designer must look for the most critical combination of loads that could occur. With inclined forms, additional horizontal load effects occur, which usually need extra bracing.

Construction Philosophy

The extent of some of these loads is indeterminate, for example, impact. If a 2500 kg skip of concrete moving at 1.0 m/s strikes a form and comes to rest in 30 mm, the force on the form will be 41.6 kN. In other cases, heavier skips, greater velocities or quicker decelerations of the load, with result in bigger impact forces. If the formwork were to be designed to resist this impact, so that it would not be distorted beyond its tolerances, it would become both prohibitively expensive and bulky.

Clearly it is not practical to design the form to fully withstand this high load. Further, it is not the only case where there can be overloads on forms. Mounding of concrete can be excessive, unexpectedly high winds can occur, flash flooding may endanger footings to the supports or incorrect load placement may create instabilities.

To effectively cater for these possibilities, however unlikely they may be, one must examine the form arrangement for an understanding of the mode of failure that might accompany an overload or combination of overloads. Both precipitate and progressive collapse must be avoided.

An example of progressive collapse was given in Figure 1.36. This showed the outer prop of a soffit form failing, and then its load transfered to the adjacent prop. The bearer then acted as a cantilever which, firstly, increased the load on this prop, and secondly, rotated the top of the prop with a bending action on it. If this second prop failed then the failure mode would be likely to progress to the next prop, hence, progressive failure.

Another example of a progressive collapse, that occurred on a building site, was initiated by one of the supports that was seated on a concrete surface near the top of a 150 mm step. A minor sideways movement in the form support caused it to drop over the step. The load on the prop was immediately transfered to the adjacent supports. A progressive collapse ensued.

A case of the precipitate collapse of incorrectly assembled frames was indicated in Figure 1.35. The collapse from overload most probably commenced similar to the relatively slow buckling failure shown in Figure 1.34. However, as the connector pins between adjacent stories of frames were not installed there came a point in the buckling failure when one frame separated from another. The collapse beyond that became precipitate and violent. Full size tests on three storey frames clearly indicate the tensile action between frames during buckling failure. The connector pins are essential.

Bearing in mind that heavy impact loading is very uncommon, it is reasonable to expect some structural failure from its effects. Nevertheless, it is vital that the form be devised so that when one part fails it will not generate falling debris or initiate progressive failure.

The Causes of Collapse

In addition to the obvious reasons of under-design for the expected loads, or the occurrence of great overloads, structural failure of the formwork can result from the way in which the loads are transmitted within the structure. These can be from the construction of the form structure with excessive eccentricities, or from the redistribution of loads that can occur if the form structure moves.

Eccentricity was illustrated a number of times in Chapter 1. Figure 1.27 showed a bearer offset on a prop cap plate. Figure 1.28 illustrated how the same effect can occur with out-of-square timbers. Props may be bent (Figure 1.29) or erected excessively out of plumb (Figure 1.30), both of these can lead to too much eccentricity. Also, the method of framing beam formwork can introduce large and dangerous bending effects in the supports (Figure 1.33).

At the bottom of the form supports, eccentricity can occur from variations in the foundation material (Figure 1.32) or uneven surfaces on concrete slabs where the support is located (Figure 1.31).

However, even with the best workmanship these eccentricities cannot be totally eliminated. They are, to some extent, an inherent feature of the basic materials used in formwork, and will also occur with normal good workmanship. Their effect can be minimised by good construction practice and an understanding of how to incorporate, in the formwork construction, the 'best' mode of structural behaviour of the support system used. This will be discussed later in this chapter.

CONVENTIONAL SOFFIT FORMS

Many completely prefabricated modular soffit systems have been developed and some of these are reviewed later in this chapter. However, the majority of soffit formwork is done with the construction of plywood, timber joists and bearers (runners) carried on steel support systems. These supports are carried on a load spreading or footing arrangement. Figure 6.02 shows such a conventional system. In this case the supports are adjustable telescopic props.

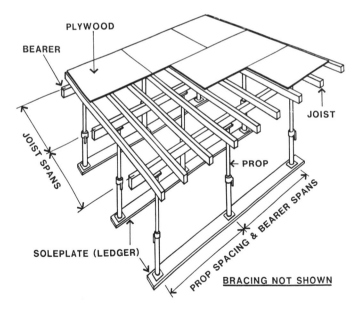

Fig 6.02 - CONVENTIONAL SOFFIT FORMWORK

The way in which the bearers apply their load to the top of the props and the arrangement of the footing system in providing a reaction to these load has a large effect on the carrying capacity of the supports. The upper framing arrangements will be discussed first and then a range of footing systems.

PLYWOOD

It is almost universal practice to construct the formface material from plywood. In the majority of cases plywood surfaced with phenolic type impregnated paper is used to control moisture absorption and the resulting hydration staining. This topic along with other pertinent matters on plywood, sealed and raw, and the use of solid timber planks in its place were discussed in Chapter 2.

Typically, plywood sheets for formwork are made with the face grain parallel to the length of the sheet and this is the direction of greatest bending strength in most cases. To minimise installation and stripping costs, reduce damage and extend the sheet life, nailing of the plywood to the joists should be kept to a minimum.

Notwithstanding this, there is a need to develop continuity in the soffit form to inhibit any tendency of the form to move apart. If the plywood sheets are placed, lapped as shown in Figure 6.03, a degree of continuity is achieved. This comes through the mat of the reinforcement which is supported on bar chairs carried on the plywood and from friction between the plywood and the tops of the joists.

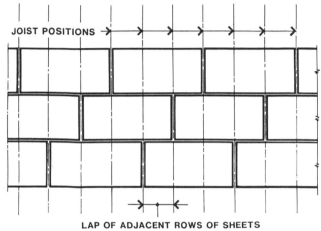

Figure 6.03 - LAPPING OF PLYWOOD SHEETS

To make the continuity effective, all of the plywood sheets on the periphery of the soffit form should be nailed to the joists that support them. This perimeter tie line is to contain the interior plywood sheets in their positions.

The plywood sheets will be structurally continuous over a number of spans (joist spacings). The maximum span will be determined by:

1. Load vs. span characteristics - Selection tables are available for this.
2. Deflection vs span characteristics - Selection tables are used. The limiting deflection is usually controlled by the specified concrete surface quality and the associated tolerances.

For efficiency, the selected span should be one where multiple spans can equal the available plywood sheet length. For a 2400 mm long plywood sheet these could be 200, 300, 400, 480 and 600 mm; for a 1800 sheet: 200, 300, 450 and 600 mm. The selected span must not be longer than the maximum span determined on strength and deflection criteria.

Other matters to be taken into account in the selection of the plywood sheet size are:

1. Length and width to give minimum sheet cutting to suit the overall dimensions of the area being formed.
2. Weight of sheet suitable for manual hoisting in the particular case.
3. Exposure to wind. Large plywood sheets are difficult and often dangerous to handle in high winds, e.g. on the upper floors of high-rise buildings.

SOFFIT FORM FRAMING

Joists

Joists for conventional formwork are typically solid softwood timber members, oregon (Douglas Fir) being one of the most commonly used timber species. They span between the bearers and carry the loads from the plywood. The materials for timber members were discussed in Chapter 2. Figure 6.04 shows a plan of a typical arrangement of plywood sheets and their supporting joists.

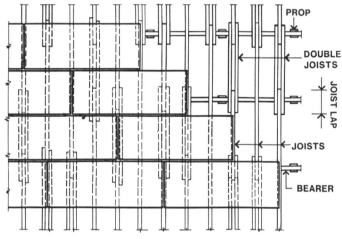

Figure 6.04 - PLAN OF PLYWOOD SHEETS AND JOISTS

Most joists are shown continuous over several spans; others are only one span long. At their junctions on a line of joists they are lapped. The total lap length is usually about one fifth of the joist span. A lap of this length is needed to ensure safety if the joist is displaced a small amount.

If this joist lap coincides with the end junction of two plywood sheets, an extra joist must be added to ensure support to both sheets. In other cases the joists can simply lap past one another. However, when the joists are being placed the formworker usually does not know the plywood pattern, so double joists are often placed at all laps.

Two situations at the end junction between plywood sheets are shown. If the sheets butt over a single joist, care must be taken to ensure that each bears half way across the joist width. For safety, some nailing is usually needed here. A small longitudinal movement of either of the plywood sheets might lead to it dropping off the edge of the joist. For the second case where the end junction of the plywood sheets is carried on two members, because it

is at a joist lap, nailing is not usually needed. Here, there is normally sufficient projection of the sheet beyond its bearing line to ensure a safe situation even if it moves a short distance.

The size of the joists is determined from considerations of material strength, deflection limitations, section stability and formface accuracy. The factors controlling strength and the need to re-assess the quality of timbers at each use were discussed in Chapter 2.

Joist deflection, like plywood deflection, is subject to tolerances related to the specified concrete surface quality and its associated accuracy.

As described in Chapter 2, the stability of the joist cross section has two aspects. The first is structural stability under load. Narrow deep timber sections, particularly where neither the top face at mid-span or the bottom face at the supports are held against sideways movement by anything more than friction, can buckle sideways when loaded. To use deep narrow timbers, the top faces must be fixed to the plywood and the bottom to the supports. This is not conducive to ease of stripping. The solution is to use sections that are proportioned so that fixing is not needed to achieve stability.

In general these are as wide as half their depth for small sizes, and not narrower than 45 mm for depths to 90 mm. Recommended timber size guidelines to achieve an acceptable level of inherent structural stability are:

(1)   Minimum width 45 mm.
(2)   For 45 mm width, the depth should not exceed 90 mm.
(3)   For 70 mm width, the depth should not exceed 190 mm.

Even when these sizes are used, there is a need to check the lateral stability of the section under the design load for each case.

The second joist stability consideration concerns worker safety. Formworkers often move about on joists and bearers prior to placement of the plywood. Joists with narrow deep cross sections can roll over quite readily when walked on. Wider sections are safer. For this reason square sections are sometimes used for joists; they have less tendency to roll over.

When first placed in position the joists often lie on their wide face. Formworker may walk on them or fall on them. They must be strong enough in this situation to carry this load when on their 'flat'.

The final consideration on joist size selection concerns the accuracy of the soffit formface. If the depths of the joists vary even a few millimetres, this will be reflected in undulations in the formed concrete face. Uniformity of joist depth is needed and the variations that occur between depths in sawn joists as they are often delivered from the sawmill are not acceptable. To achieve uniformity, it is common practice for joists to be 'sized' down 3, 4 or even 5 mm in depth; to 97 mm for 100mm sections, 147 mm for 150 mm sections etc. The reduced sizes of the joists must be taken account of in the design calculations.

Even if the most stable of cross-sections are used for the joists there will often be a need for some of the joists to be fixed to the bearers that they sit on. The extent of these requirements will depend on the support system used. If the supports are a fully braced and tied modular system then joist fixings may only have to be minimal.

However, if the supports are telescopic props then there will be a need to stabilize the top of the props by fixing them to the bearers. In turn, the bearers must be fixed to some of the joists and, at least, the plywood sheets on the perimeter of the form nailed to the joists. Joist fixing to the bearers is usually done by skew nailing.

Bearers.

Bearers, sometimes called runners, are the more heavily loaded of the soffit form framing members and may be large solid timber sections, laminated timber beams, manufactured timber I sections (Figure 6.05) or metal beams.

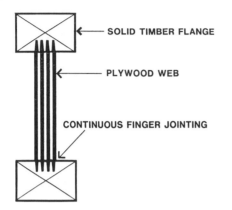

Figure 6.05 - MANUFACTURED TIMBER I BEAM

This description of the general principles of bearers will be confined to solid timber sections. These solid sections range in width from 70 mm to 100 mm wide. The 'U' heads and flat cap plates of many proprietary support systems are 200 mm square to permit the lapping of these sizes of bearers over the support.

As bearers are heavily loaded, and usually long in span, deflection limitations often control the selection of the bearer size. For this reason it is common practice to use bearer members long enough to be continuous over at least two spans. This has the benefit of reducing deflection to less than that which would happen with a single span bearer of the same size.

The detrimental effect of eccentricity of load application on the structural capacity of the support systems has been noted previously. Where the formwork area is narrow in width and the involves only one double span bearer, the achievement of general concentric action is easily done. (Figure 6.06)

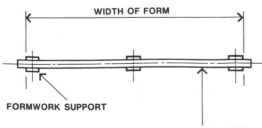

Figure 6.06 - DOUBLE SPAN BEARER

Wider formwork areas will need a line of double span bearers. If they are lapped as shown in Figure 6.07-A, the support in the middle of each double bearer span will have the **full load acting eccentrically**.

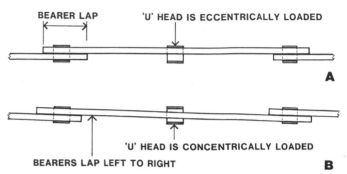

Figure 6.07 - THE LAPPING OF BEARERS

The staggered laps shown in Figure 6.07-B effectively eliminate this mid-length eccentricity. In both cases the laps between adjacent bearers cause each of the bearers to act eccentrically. However, the effective eccentric load bending the support head cannot ever be more than one half of the load on the support at the mid-length of the double bearer. This will only occur when one bearer is fully loaded and the adjacent one is not loaded.

Where there is only one bearer on a support, or the double bearers do not totally fill the available width of the 'U' head, the bearers may move sideways on the head during formwork construction or concrete placing. This movement can be inhibited, and general concentricity maintained, by rotating the 'U' head as shown in Figure 6.08. To prevent it drifting back during construction it should be fixed with a nail through at least one of the holes provided in its sides.

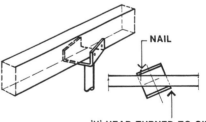

Figure 6.08 - CENTRALISING A SINGLE BEARER

Alternatively, the space between the sides of the bearer and the inside of the 'U' head can be packed to hold the bearer concentric. The packing timbers should be nailed to prevent their dislodgement.

If only single span bearers are used, the staggered laps shown in Figure 6.09 can be a disadvantage. The problem occurs when the progress of the slab pour is on a line at right angles to the bearer line. As the pour progresses, the initial eccentricity on each support is always acting in the same direction as that which occurred initially on the previously loaded support. The progress of the pour past each support generally balances out the eccentric moments, as the full load is reached on each support. Nevertheless, there is an overturning moment continuously acting the one way, at the edge of the progressing pour. This has the potential, together with other influences, to destabilise the formwork structure.

PROGRESS OF CONCRETE PLACEMENT

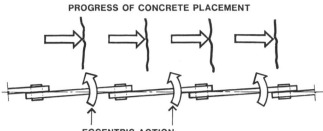

ECCENTRIC ACTION

Figure 6.09 - BEARERS THAT LAP THE SAME WAY

PROGRESS OF CONCRETE PLACEMENT

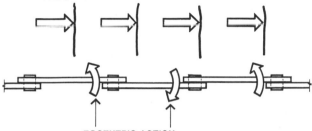

ECCENTRIC ACTION

Figure 6.10 - BEARERS LAPPING ALTERNATE WAYS

The danger can be minimised by lapping the single span bearers as shown in Figure 6.10. In this arrangement each succeeding bearer causes an initial eccentric moment on the support which acts the opposite way to that which acted on the previously loaded one. This method of bearer framing has less risk than the previous one.

Where the progressive line of the edge of the pour is parallel to the bearer line, the growth in loading on all bearers is generally uniform, and neither method gives any particular advantage over the other.

The importance of the control of deflection of the bearers was noted previously. To this must be added considerations of material strength and section stability in the determination of size. The factors that control material strength and the need to re-assess the condition of the timmber at each and every use were given in Chapter 2 and discussed earlier for joists.

The information on joists, given previously, included recommendations on selecting section dimensions for best stability under load. The sizes typically used for bearers, depending upon loading and span, are: 140 mm deep x 70 mm wide, 150 mm x 75 mm, 140 x 90, 150 x 100, 200 x 75 and 200 x 100.

Where the head of the support is fitted with a 'U' head it is convenient to select section widths such that two bearers will neatly fill it (e.g. 2/100 mm wide bearers for a 200 mm wide 'U' head). This avoids the matter of rotating the 'U' head, or packing it at the sides to stabilise the bearers at the support.

If the top of the support is a flat plate, the combined width of the two bearers should not exceed the plate width. The bearers should not overhang the plate. For a 150 mm square cap plate the bearer widths would be limited to 2/75 or 2/70 mm wide bearers.

For accuracy in the line of the formed face, 'sizing' of the bearers is not normally required; variations in the depths of nominally equal sized bearers of up to 5 mm are common. The important matter is the accuracy and straightness of the line of the top of the bearers. If there are differences in the depths of adjacent bearers the shallower bearer is packed to give a uniform top line. (Figure 6.11) The adjustment devices in the support system are then used to adjust the whole form to have the required surface level.

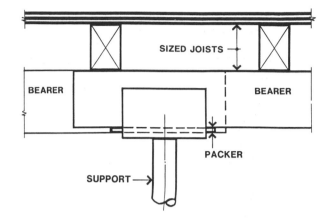

SIZED JOISTS

BEARER   BEARER

PACKER

SUPPORT

Figure 6.11 - PACKING THE BEARERS TO LEVEL

FOOTINGS AND FOUNDATIONS

The foundation can be defined as the earth surface which carries the load. The footings are the part of the formwork structure that is in direct contact with and transmitting the load to the supporting foundation. The footings will not always be spreader timbers at the base of the supports. In some cases the footings will be a reinforced slab on ground, in others it may be the footings of the permanent structure being built, where the formwork in question is being constructed on some upper level suspended slab. They can even be timber piles driven solely for the formwork.

Some parameters can be given on the characteristics of a satisfactory footing system and its supporting foundation. It must be able to adequately resist the vertical forces, horizontal forces, eccentricities of action of the forces and the influences tending to move the formwork structure. If vertical movement occurs it may affect only individual supports or alternatively, every support might sink. In both cases the tolerances for the final concrete structure could be exceeded.

In this matter, plywood, joist and bearer deflections are not a consideration; they relate only to relative surface alignment and the visual quality of the concrete.

The movements in question here are those that affect the whole or significant parts of the structure. They may occur before or after the initial set of the concrete. Movements that occur prior to the initial concrete set, can usually be corrected by adjustment of the formwork. Any movement after the initial set is usually detrimental to the concrete structure. It will only be acceptable with the permission of the engineer responsible for the supervision of the construction of the concrete structure. It is rare for it to be permitted.

If experience, foundation material tests, or structural analysis of the formwork indicate that there will be movement under load, the height and position of the formwork can adjusted beforehand, e.g. camber in long span slab forms, or to cater for vertical settlement of soft soils.

Three general categories of bearing situations can be considered: bearing on the suspended structure of the previously built work, bearing on reinforced concrete slabs on the ground, and being supported on temporary structural elements. The last case may be simple timber ledgers or, in extreme cases, temporary piling.

If the supports are seated on a suspended concrete structure the loads on the form will cause some deflection of the structure. This deflection may be insignificant or it may be quite large. (Figure 6.12). Two considerations arise from the latter case: catering for the effect of this deflection on the accuracy of the new work, and reducing the load effect on the supporting structure to acceptable levels.

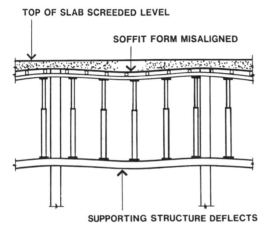

Figure 6.12 - PERMANENT STRUCTURE DEFLECTING

The supporting structure deflects under the load and the soffit formwork follows that shape. The top of the concrete is screeded level with a resulting increase in concrete depth and weight. To avoid this, the soffit form can be cambered up to counter this anticipated deflection.(Figure 6.13)

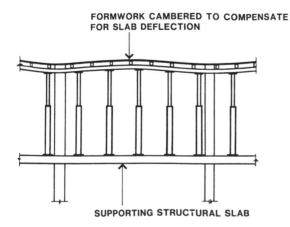

Figure 6.13 - SOFFIT FORMWORK CAMBERED

It should be noted that the additional load of the fluid concrete will cause some further creep deflection of this supporting slab. However, this is not a consideration for the formworker; it is for the design engineer for the permanent structure to determine this and decide what action is to be taken.

To minimise the load on the supporting slab and reduce this deflection, the load from the new work can shared by other suspended slabs below as shown in Figure 6.14.

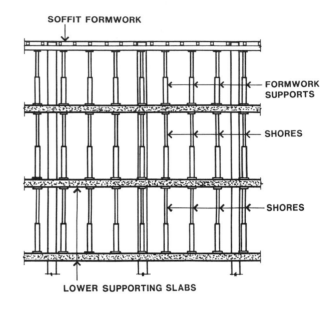

Figure 6.14 - THREE LEVEL MULTI-STOREY SHORING

The shores at the lower levels should closely align with the positions of the formwork supports. This eliminates this source of secondary bending effects on the slabs,

Soffit supports are often seated on reinforced concrete slabs on the ground. Its settlement response to the point loads will depend on the soil sub-grade characteristics and compaction, slab thickness and reinforcement, and concrete quality. Some vertical movement will always occur; it has been regularly observed to range from insignificant amounts up to 5 mm. The movement is usually uniform for uniform support loads and it is normal practice to set the form at a slightly higher level than specified, to cater for the anticipated settlement.

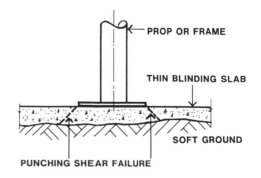

Figure 6.15 - POSSIBLE FAILURE OF A BLINDING SLAB

The predictable behaviour of properly constructed reinforced concrete slabs cannot be applied to the responses of thin unreinforced blinding slabs. These slabs are a quite common feature of the first stage of the construction of sandwich floor slabs. As Figure 6.15 indicates, punching shear failure is possible for blinding slabs on top of soft ground. Formwork supports must not be placed directly on such slabs. Substantial spreader timbers should be placed on the slab (Figure 6.16) to increase the bearing area and minimise the chances of punching shear failure.

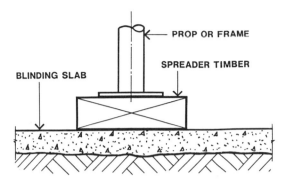

Figure 6.16   SPREADER TIMBER UNDER PROPS

Although most concrete slabs are intended to have smooth even surfaces, variations in the concrete surfaces do occur. These variations can cause eccentric loading and limited rotation of the base plate to the formwork support. This situation is illustrated in Figure 6.17.

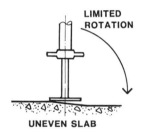

Figure 6.17 - EFFECT OF AN UNEVEN CONCRETE SLAB

If uneven patches like this are prevalent on the concrete surface, then soleplates under the baseplates will be needed to provide an even level bearing.

Direct bearing on the ground can pose a range of problems. Even if the bearing capacity of the soil's surface appears good, it must be remembered that construction sites are usually very disturbed. Drainage is often installed before suspended slabs are formed, and the backfilling to the drain trenches is rarely fully compacted. Further, this backfilling can often contain brickbats and rocks. If a soleplate bears on this combination of soft soil and hard points, as shown in Figure 6.18, there will be eccentric loading of the support baseplate, with no effective limitation on the rotation of the baseplate.

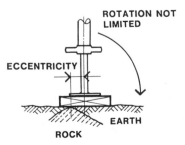

Figure 6.18 - EFFECT OF UNEVEN GROUND

Poor compaction of the backfill to strip footings can cause a similar problem for formwork supports that are adjacent to walls. (Figure 6.19)

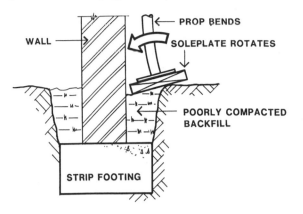

Figure 6.19 - PROP PLACED ON TRENCH BACKFILL

These problems of variations in bearing capacity can occur in all soils and all construction sites. As well as the soft areas that can come from drainage and footing work, the surface area as a whole may be disturbed by the traffic of delivery vehicles and earthmoving plant. Proof rolling of the surface may be necessary in this case.

Further, the natural (or excavated) surface material may lack bearing capacity. Soils that appear to be adequate when dry, may be grossly deficient in bearing strength when wet. Rain, water from other construction work and leakage from the concrete pour can always occur; the soil bearing capacity must always be evaluated for the wet condition. This evaluation to determine the allowable bearing pressure for footing design, usually requires a professional geotechnical investigation.

If a simple sole plate arrangement, such as a scaffolding plank, is used on soft soil, the combined effect of its flexibility and the soft surface may lead to the unsatisfactory situation shown in Figure 6.20.

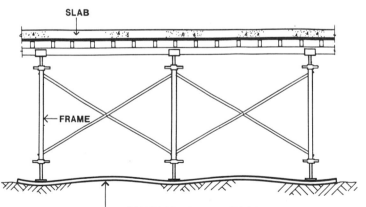

Figure 6.20 - FLEXURE OF A THIN SOLE PLATE

Here the formwork support is only being effectively carried by the small portion of the soleplate that is near the support. Such a small foundation area results in high bearing pressures that exceed the bearing capacity of the soft soil. As an inevitable consequence, the formwork sinks when loaded.

To enlarge the foundation area, reduce the bearing pressures and minimise sinking, the soleplate must be made stiffer. This can be done by adding members to the top of the soleplate. To be effective they will also need to be stiff. Members similar in size to the bearers used in the soffit framing are usually suitable. However, if only one is used there are potential stability problems. The situation shown in Figure 6.21 could become unstable.

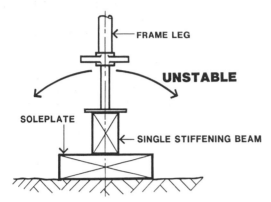

Figure 6.21 - SOLE PLATE WITH SINGLE STIFFENER

As a good general rule, all framing arrangement should be examined to see what would happen if the relative position of the various components were shifted slightly. Would instability result? In the case shown a minor shift sideways of the baseplate on the single stiffening beam would produce eccentricity and the danger of this beam rolling over. A more stable stiffening configuration is needed.

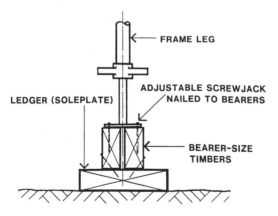

Figure 6.22 - SOLE PLATE WITH DOUBLE STIFFENER

As Figure 6.22 shows, the possibility of relative movement is reduced by fixing the baseplate to the stiffening timbers. The use of two or more stiffening beams provides width of bearing, and greater stability results. Note also that the soleplate must bear evenly on the ground. Care must be taken to trim the bearing suface accurately.

It is not acceptable to seat the soleplates on a sand layer to correct innaccurate surface preparation. This is dangerous as sand is readily liquified by water.

If the bearing capacity of the ground's surface is so low that even this arrangement will not obtain bearing stresses within its capacity, then wider bearing (and low pressures) can be achieved as shown Figure 6.23.

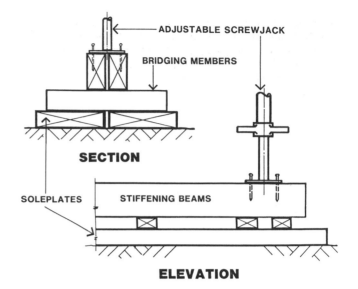

Figure 6.23 - WIDER BEARING AREA FOR SOLE PLATE

In circumstances where it is considered that there is the strong possibility of movement of the support baseplate, perhaps by impact, the detail shown in Figure 6.24 can be adopted.

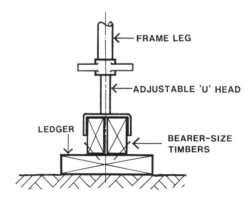

Figure 6.24 - 'U' HEAD AS BASE PLATE ON BEAMS

Instead of the normal flat baseplate adjustable screwjack being used at the bottom of the support, a 'U' head is used upside down. It cannot move sideways on the double bearer-size timber beams.

As implied in the earlier descriptions, the baseplates to the supports should never be directly supported on soft soil surfaces. The bearing area is simply too small. For a quite different reason they should never be founded directly on hard or rocky surfaces. In these cases the uneven nature of the surface can lead to unacceptable eccentric action on the formwork supports.

If the surface cannot be cut to an even bearing line then suitable packing to fit the rough surface must be provided. Two features are needed: the packing must be waterproof and there must be a means of holding the baseplate in position against any minor movements or impact. A suitable method is shown in Figure 6.25.

A hessian (burlap) bag of lean-mix, and relatively dry, concrete is placed directly under each formwork support. A soleplate, usually a timber scaffold plank, is placed along a line of supports and forced and packed down to give a good bearing of the concrete to the rough ground, and the concrete to the underside of the plank. The supports are erected on and fixed to the plank, centrally above the concrete pads, to transmit their loads directly to the ground. As indicated in the Elevation of Figure 6.25, the flexural strength of the plank is not involved in the load path.

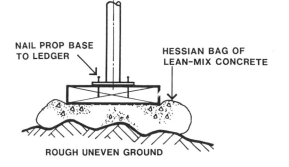

SECTION

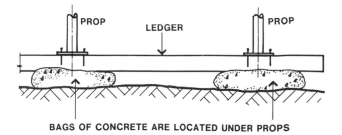

ELEVATION

Figure 6.25 - CONCRETE PADS UNDER THE SUPPORTS

If the geotechnical investigation shows the surface bearing capacity to be grossly inadequate, and it is not possible to provide adequate formwork footings by wider bearing areas or improve the capacity by proof rolling then more expensive footing techniques will be needed. These can be discussed under three headings: **Ground Improvement, Special Footings** and **Use of the Building Structure**. The following description of these is not intended to be either complete or exhaustive. It merely provides an overview of this topic.

Ground Improvement.

Poorly compacted ground is often open in texture and subject to relatively large settlements when loaded. Two types of ground improvement can be used. The first is grout injection. The grout can be either cement/sand grout or lime/sand grout. The latter, although lower in strength, has the advantage that it does not impede any later excavation work. This is compatible with a general principle that the remnants of any temporary construction techniques should not be a future problem for the building owner.

The site access required is minimal, as the grout can be pumped over a distance from the mixing point, to probes inserted in the ground. The holes for the probes can usually be drilled by hand-held equipment.

The second general method of ground improvement involves improvement of compaction. This involves the insertion of high powered vibratory probes into the ground and is usually only effective with granular material such as sand. A high ground moisture content facilitates the fluidization and compaction. This method requires expert geotechnical guidance and full site access for the large equipment involved.

Special Footings.

In very poor ground, such as waste product land-fill sites or alluvial areas, a suitable bearing material may be a considerable depth below the construction site surface. One solution to this problem is temporary construction piling.

Where the founding strata is deep, which is very common in bridge construction, steel I sections are often the best solution. They can have a high load capacity and can usually be withdrawn after the construction work is complete. Smaller section steel piles for relatively light loads can often be driven by attachments on hand held pneumatic equipment.

Concrete piles, either precast or in-situ steel tubed piles can be used. However, they are usually be very difficult to remove.

For less than 20 tonne capacity and depths less than 8 metres, timber piles of 150 mm to 180 mm diameter, are often the most economical solution. However, the problem of attaching a tensile fastening to the top often makes them difficult to withdraw. For construction piling to be left in place after the completion of the work usually requires the permission of the property owner.

Use of the Building Structure for Footings.

On sites where the soil surface has insufficient bearing capacity to carry the formwork, it is common for the permanent structure, building or bridge, to also require deep foundations such as piles.

Rather than install expensive temporary piling, one solution to the problem is to support long span formwork, such as table forms, on the building foundations. Figure 6.26 diagramatically shows an arrangement of this type.

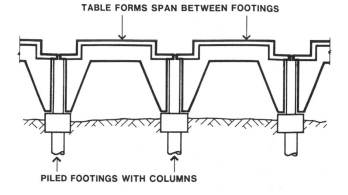

Figure 6.26 - FORMWORK SEATED ON THE BUILDING FOUNDATIONS

The figure shows the formwork spanning between the pilecaps of the permanent structure. If these forms are made as a large single unit, a tableform, a means of lowering and withdrawing it in one unit also has to be provided. A temporary road may have to be made on the soft ground.

SUPPORT SYSTEMS

As described previously, the support members of the formwork structure must always be regarded as subject to eccentricity of load action at the top and and eccentricity of reaction at the bottom. The extent of the eccentric action will depend on the direction of the framing members. For example, if a support is seated on a stiffened sole plate as shown in Figure 6.27 its baseplate can be subject to rotation in both directions, labelled 'A' and 'B'.

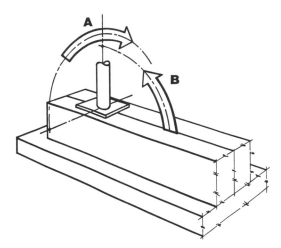

Figure 6.27 - POSSIBLE ROTATION OF BASEPLATE

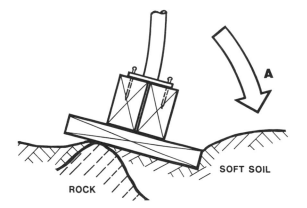

Figure 6.28 - LATERAL ROTATION OF BASEPLATE

Figure 6.28 shows the lateral rotation, in the direction 'A', of the base of the support in response to a hard point of bearing in a generally soft area. The light steel tubular support is the only thing endeavouring to inhibit this rotation. For safety reasons, all but the hardest soils must be regarded as soft when wet, and offering little restraint to this rotation.

In contrast, in the direction 'B', which is along the line of the soleplate, the rotation of the baseplate is very effectively limited by the combined stiffness of the light tubular support and the heavy timber soleplate. This is illustrated in Figure 6.29.

Not all support systems have equal stiffness in both directions. Prop systems and modular systems do; frame systems are one example of unequal stiffness. Knowledge of how the direction of greatest eccentric moment is controlled by the orientations of soleplate and bearer

framing can be used to direct the bottom and top eccentric loading effect in the strongest direction. This topic will be discussed in more detail with 'Frame Systems'.

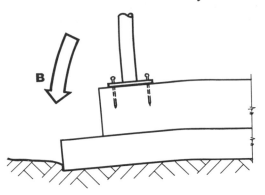

Figure 6.29 - LONGITUDINAL ROTATION OF BASEPLATE

All support systems should have a means to prevent dislodgement of the supports if one end of the bearer should lift as shown in Figure 1.07. Further, all components of the system should be able to cater for, or resist any tendency to move in relation to one another. The influences promoting these movements can be vibration, minor impact or the responses to the expected loadings.

All supports must have at least one vertical adjustment device; frame and modular systems routinely have two, one top and one bottom. This adjustment device may be called upon to fill three roles.

Firstly, the adjustment is used to bring the soffit form to the correct level and the bottom adjustments will be used to cater for variations in the levels of the footing areas. Secondly, during concrete placing, minor movements may occur and these can be corrected with the adjustment devices. Finally, the adjustment devices are used to lower the formwork for stripping. For this final purpose it is imperative that neither of the first two adjustment activities reduce the remaining available adjustment travel to less than 25 mm. Ideally, there should be at least 50 mm available down travel for stripping.

The support system and, indeed, the whole forwork structure must be safe and stable at all times, and provided with a means for safe access for formworkers. Stability calls for bracing systems. These may be lightweight erection bracing provided as part of the systems (e.g. Frame Systems) or extra temporary braces which may be removed after the completion of the final bracing.

The final bracing may be provided within the system, or be additional tube bracing and horizontal ties added to the formwork supports. The ability of framed bracing to resist horizontal forces depends on the total loads acting on the supports incorporated in the bracing. If this is only the weight of the formwork and reinforcement there will be little resistance to overturning the braced bays. (Refer to Figure 1.05) If, however, the concrete is in place on the form, this weight is greatly increased and the resistance to overturning also increases proportionately. For this reason bracing should be located under the formwork areas where the concrete placement will commence.

Unfortunately, in most cases the formworker will not know where the concrete pour will start and will have no authority to dictate the start point. As concrete pours usually start at an edge of the pour it follows that the required bracing should be installed uniformly around the perimeter of the pour. This will maximise the probability that the first loads of concrete will be placed over or very near to a bracing set.

One common bracing method is to connect the system to the previously cast columns or walls. Because of the potential for structural damage, this requires the permission of the engineer for the permanent structure.

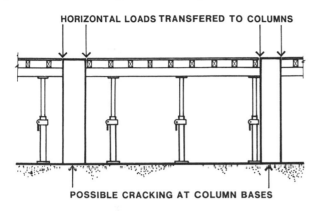

Figure 6.30 - BRACING TO THE PERMANENT STRUCTURE

Figure 6.30 illustrates the great bending action that occurs to the columns when a load is applied to their top level. If this load is impact from a swinging concrete bucket, bending failure of the column seems certain for all but the largest and most heavily reinforced columns.

****

A large number of proprietary systems are available. They range from the simple tube, coupler and screwjack combinations to sophisticated modular systems. This section will be confined to general details of Prop Systems, both tube and telescopic, Frame Systems and Modular Systems.

Prop Systems.

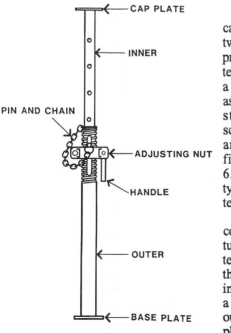

Figure 6.31 - TELESCOPIC PROP

In general there can be said to be two categories of props: proprietary telescopic props, and props assembled from standard steel scaffolding tube and proprietary fittings. Figure 6.31 shows a typical proprietary telescopic prop.

The device consists of two tubes, the inner telescoping within the outer. The inner is fitted with a cap plate and the outer with a base plate.

At the top of the outer a square-form screw thread has a vertical slot cut on both sides. A large steel nut with a hinged handle can be wound up and down this thread. A hardened steel pin, retained by a chain to a ring over the inner, is inserted through the slots and the regularly spaced holes in the inner, just above the nut.

Sometimes these pins are lost, and it has become a common and dangerous practice to substitute an off-cut of reinforcement for the pin. This is a safety hazard, only the hardened steel pin should be used.

Coarse adjustment of the prop length is done by removing the pin, sliding the inner up to the required position and re-inserting the pin. Small adjustments are made by winding the nut. This adjustment method is the most common type. However, there are variations in the adjustment and pinning mechanism on some brands.

These telescopic props are available in a number of size ranges. A typical brand has its shortest at 1.09m closed/1.96m open through a total of five sizes up to 3.28m closed/5.13m open.

Some suppliers also market heavy duty adjustable props. These range from aluminium telescopic props to steel segmental framed props.

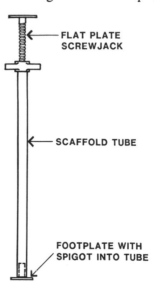

Figure 6.32 - TUBE PROP

As noted above, the second category of props consist of fittings and standard steel scaffold tube (normally 48.4 mm O.D. x 0.4 mm wall tube). A screwjack is inserted into the top. The shaft of the screwjack is normally a neat sliding fit in the scaffold tube. At the base there may be another screwjack or, as shown in Figure 6.32, a simple footplate that has a short spigot which is inserted into the scaffold tube.

Compared to proprietary telescopic props these scaffold tubes are quite slender.

If they are taller than 2.0 m, mid-height horizontal ties in two directions at rightangles are usually needed. These ties effectively reduce the buckling height of the props and increase their load capacity. Standard scaffold tubes and couplers are used for these ties and they must all be connected into bracing sets.

Very tall proprietary telescopic steel props can also get structural benefit from tying at the mid-height. Reference should be made to the manufacturer's technical literature for the requirements and the load carrying benefits to be had from this procedure.

One of the major disadvantages of props as a formwork support system is their individual instability until they are connected into the total form system of props, framing and bracing. The props have to be held in place, usually manually, until either the tube ties or the soffit framing are connected to them.

Some brands of props are fitted with retractable braced feet to give the free standing prop some individual stability. (Figure 6.33)

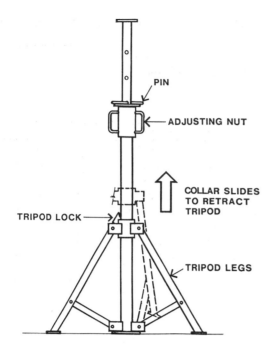

Figure 6.33 - PROP WITH RETRACTABLE TRIPOD

A similar device is a separate bracing tripod (Figure 6.34) which can be temporarily fitted to the prop to hold it vertical until the bearers, joists plywood formface and final bracing are fitted. It fits around the shaft of the prop and is held in position with captive wedges. Once the prop is tied or braced the tripod can be removed.

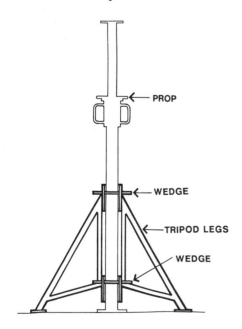

Figure 6.34 - PROP WITH REMOVABLE TRIPOD

At the least, the fixing of the bearers to the prop cap plate should be two 65 mm nails. (Figure 6.35) If the bearers cannot be nailed, for example when using aluminium beams as bearers, then the props must be tied together with scaffold tubes near the top. In this case it is normal and safe practice to leave this temporary bracing in place until at least the day after the completion of the concrete pour.

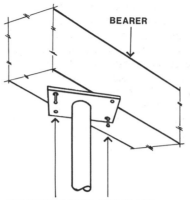

Figure 6.35 - FIXING BEARER TO PROP

As noted previously, eccentricity of load action can quite readily occur at both the cap-plate and the base plate of the props. Perfection in the accuracy of erection of props and framing members is simply not realistic. Formworkers must develop an understanding of what is a sensible dividing line between normal trade practice and negligent workmanship. Two good standards to adopt are: maximum top eccentricity 20 mm and maximum out of plumb 1 in 200 and 40 mm. (Figure 6.36)

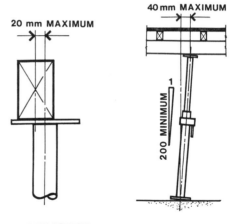

Figure 6.36 - LIMITS OF GOOD WORKMANSHIP

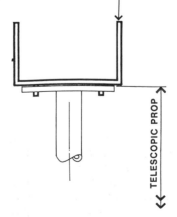

Figure 6.37 - 'U' HEADS FOR TELESCOPIC PROPS

The problems of maintaining the concentric placing of bearers over props are exacerbated by their flat cap plates. To facilitate easier control of concentricity, 'U' heads can be fitted to the props. As shown in Figure 6.37 shows, these 'U' heads are provided with short pegs that insert into holes provided in the cap plate.

To rotate the 'U' head to centralise one or a pair of bearers, the whole prop must be rotated. For scaffold tube props two devices are available. The first is the adjustable 'U' head (forkhead) which is inserted into the top of the tube. (Figure 6.38) This is described in more detail with Frame Systems.

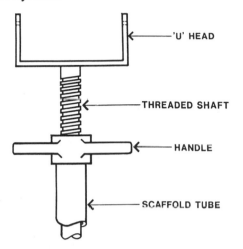

Figure 6.38 - 'U' HEAD OR FORKHEAD

The second device is a special cap plate that is fitted with a vertical fin plate to locate the single bearer concentrically. Nail holes are provided for fixing the bearer in position. As shown in Figure 6.39 it connects to the scaffold tube prop by a spigot that inserts into the top. As the top of the tube prop has no adjustment in this case, the bottom must be provided with a screwjack.

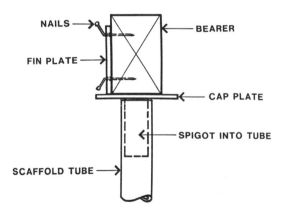

Figure 6.39 - CAP WITH VERTICAL FIN PLATE

Bracing of Props.

The general requirements for the bracing of support systems were discussed earlier in this chapter. The particular bracing requirements for props are outlined here.

There are no standard bracing components provided as part of prop support systems. Bracing sets are framed from the props, scaffold tube and couplers. Figure 6.40 shows the elevation of a bracing set. The scaffold tubes are used

for the top and bottom horizontal tube and the diagonal brace.

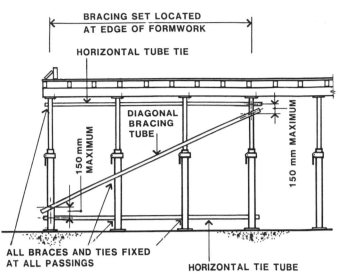

Figure 6.40 - BRACING SET FOR PROPS

Couplers are required at all passings: tube to tube and tube to props. Ideally, horizontal tubes and braces should coincide at their connections to a prop. Because of the couplers this is not possible. However, they should couple to the prop within 150 mm of each other.

This bracing set is a rigid frame and will counter the horizontal forces, applied at its top level, by its resistance to overturning (Figure 1.05) and sliding (Figure 1.03). As discussed previously, this resistance will depend on the total weight acting on the bracing set. To maximise this, the bracing sets should be located at the perimeter of the form where the concrete placement will usually commence. Figure 6.41 diagramatically shows the general arrangement of the bracing sets to comply with this strategy.

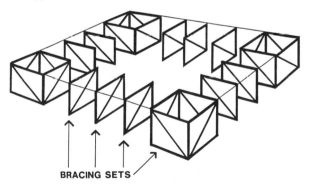

Figure 6.41 - LOCATION OF BRACING SETS TO PROPS

Propped Formwork Erection Procedures

To illustrate the precautions necessary for the erection of a simple rectangular soffit form, one method of erection will be described. The example will be assumed to be an independent form not connected to or braced by any previously constructed columns or walls. In this case, the props used will not be equipped with the bracing tripods shown in Figures 6.33 and 6.34 above.

As it is vital that the work must be stable at all times, it is necessary to commence with a braced core. This start point may be located anywhere in the formed area. However, the desirability of locating the bracing at the perimeter was discussed above, and the braced core will be located at a corner.

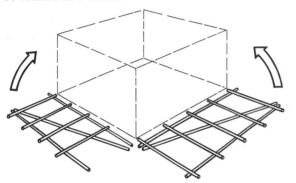

BRACING SETS ASSEMBLED AND THEN HOISTED

Figure 6.42 - ASSEMBLY & ERECTION OF BRACING SETS

Figure 6.42 shows the start of the braced core. It will consist of four braced sets. Assembly begins on the ground using props, horizontal ties and braces. The units are then lifted up to the vertical, either man-handled or by crane, and coupled together. Note that the adjacent braced sets are shown sharing a prop at each corner.

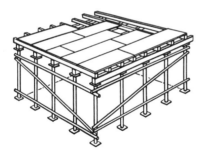

Figure 6.43 - BRACED CORE WITH SOFFIT FORMWORK

To complete the stability of the braced core, the other two sides, its internal props and all its bearers and joists should now be installed and fixed together with the plywood form deck as previously described. (Figure 6.43) This fixing work includes fixing of the plywood sheets around the perimeter of the form. This completed braced core often becomes a storage area for bundles of plywood sheets which can be progressively placed on the completed formwork as it extends out from the braced core.

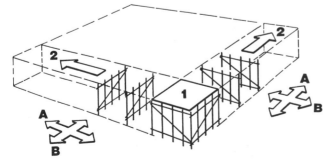

Figure 6.44 - EXTENDING OUT FROM THE BRACED CORE

The next stage is to extend 'fingers' out along the sides of the form as illustrated in Figure 6.44. Each 'finger' is continuously braced in direction 'A' by the braced core but to achieve stability in the direction 'B' it is essential that the bracing sets be completed as the work progresses. The tops of individual props and bracing sets are held by a combination of bearers and joists progressively fixed in position. As this work procedes the plywood surface can be extended out from the top.

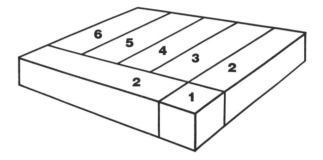

Figure 6.45 - SEQUENCE OF COMPLETING FORMWORK

The completion of these two fingers gives a stable base for the progressive completion of the remaining areas of the form. (Figure 6.45) Each of these sections must be fitted with its bracing sets as it is constructed.

Because the 'fingers' are progressively stable it is possible to expedite the completion of the total form. A number of infill areas can be constructed simultaneously, provided the adjacent areas have extended beyond them and are adequately braced. This is illustrated in Figure 6.46.

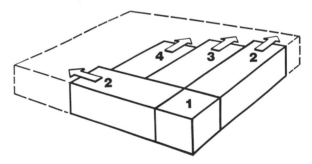

Figure 6.46 - ALTERNATIVE COMPLETION SEQUENCE

This description of an example of formwork erection should not be interpreted as a hard-and-fast set of directions. Many variations are possible. The essential point is that all completed work, partially completed work and erected components must be safe and stable at all times.

The use of previously constructed columns or walls to provide bracing has been mentioned previously. So that this bracing action can be effective, the formwork must fit firmly to the column faces.

Figure 6.47 shows the bearer and joist framing plans with a suitable plywood layout. The joist plan shows two extra joists to provide support to the ends of the plywood sheets at the column faces. To ensure that a tight interface is maintained between the column faces and the plywood ends, all adjacent plywood edges should be nailed to these extra joists as shown. The horizontal forces on the form can then be transmitted to the column by the edges of the plywood.

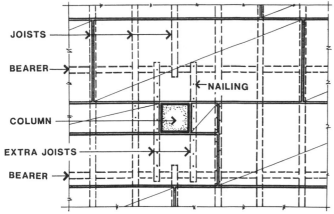

Figure 6.47 - EFFECTIVE FIXING TO THE COLUMN FACES

Figure 5.62, in Chapter 5, indicates the penetration of the column into the plywood so that this bracing connection can be efficiently achieved.

Frame Systems.

Unlike prop systems, frames are assembled in braced pairs and in general, do not have stability problems during erection. Figure 6.48 shows a typical proprietary frame assembly. This is a single storey assembly. Multi-storey frame assemblies can be used for tall form supports and their requirements and restrictions are discussed later in this chapter.

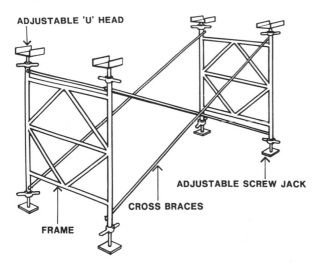

Figure 6.48 - A PROPRIETARY FRAME SET

The principal load carrying units are pairs of braced tubular frames connected together on both sides with lightweight bracing. If there is an odd number of frames in a row this extra frame is braced onto the adjacent pair. This bracing is vital for stable erection of the frames, but its primary structural purpose is to inhibit lateral buckling of the frame.

Inserted into the base of each leg is an adjustable flat plate screwjack. If adjustment is not required at the base then a simple footplate with spigot may be used. At the tops adjustable 'U' head (fork head) fitting for the support

of the bearers are shown. If, in particular cases, the 'U' heads are unsuitable then flat plate screwjacks may be used.

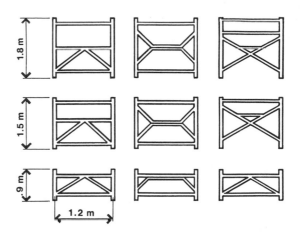

Figure 6.49 - TYPICAL TYPES AND SIZES OF FRAMES

Most of the available frame systems are approximately 1.2 m between the centrelines of their tubular uprights. They are available in a range of heights, e,g, approximately 1.8 m, 1.5 m. 1.2 m and 0.9 m. Figure 6.49 illustrates a range of frame heights for three types of frames. In most cases the frame is designed to carry scaffolding planks on its cross members at its mid-height.

Erection Bracing to Frames.

The distance apart of the frames is controlled by the light-weight cross braces shown above, in Figure 6.48. These are usually made of light tubing flattened at the ends or from angles.

The two braces in each set are permanently connected at their mid-points with either a bolt or pin. The braces are available to fix the frame spacing at a range of positions. Some manufacturers supply them for frame spacings of approximately 3.0 m, 2.7 m, 2.4 m, 1.8 m, 1.5 m and 1.2 m. Such a wide range of frame spacings are not universally available and many suppliers limit their range to 2.4 m, 1.8 m and 1.2 m.

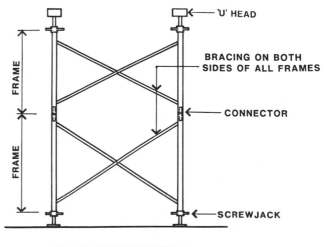

SIDE ELEVATION

Figure 6.50 - TWO STOREY BRACED FRAME SET

Every pair of frames must be braced both sides at every level. Figure 6.50 shows a two storey frame set correctly braced. It is dangerous practice to omit any of these braces. They provide buckling restraint to the frames. Without them the load carrying capacity of the frame is drastically reduced. At each junction between upper and lower frame legs, there will be a connector inserted into both. As discussed earlier, it is vital that these connectors be pinned or bolted to both the frame legs that they are inserted into.

There are variations between the various brands of devices that connect the braces to the frame. The connection sequence of a common type is shown in Figure 6.51. Note that for safety reasons, the connector is fitted to the inside of the frame leg.

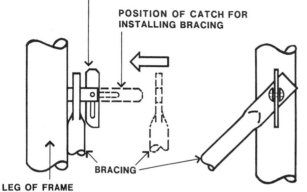

CATCH IN LOCK POSITION

POSITION OF CATCH FOR INSTALLING BRACING

BRACING

LEG OF FRAME

FRONT ELEVATION   SIDE ELEVATION

Figure 6.51 - DETAILS OF A BRACING CONNECTOR

From the illustration of the connector, it will be obvious that the frames must always be erected the right way up. If they are erected upside-down the catch will swing down to the horizontal position and the brace will readily slip off. To do this is to court disaster.

Fittings to Frames.

As shown above, the braced frames have base plates at their footing level, and bearer seating components at the top. Figure 6.52 shows a typical 'U' headed screwjack (forkhead) for the top of the frames.

It is common for the 'U' head to be nominally 200 mm x 200 mm in area, although some brands are 150 x 150. The 200 mm width readily accommodates two 100 mm wide bearers. The base area and side plates are provided with small holes (usually 4 mm) so that nails can be driven to retain the timbers. If 'U' heads are not suitable, flat headed screw jacks can be used. The general features of these are similar to 'U' headed screwjacks.

The stem of the unit may be a solid shaft or thick walled tube. It will be machined to have a screw thread, usually of generally square form. A heavy duty metal handle (wingnut) runs on this thread and is provided with a machined recess on its underside to ensure accurate location of the unit on the vertical leg of the frame.

Rotation of the handle controls the height of the 'U' head and some can extend 500 mm above the top of the frame leg. Naturally, the load capacity reduces as the extension increases. To ensure stability a minimum

insertion of the 'U' head shaft in the frame leg must be maintained. This is controlled by a stop on the screw thread, usually a small bead of weld. The stop effectively prevents the handle being wound down too far.

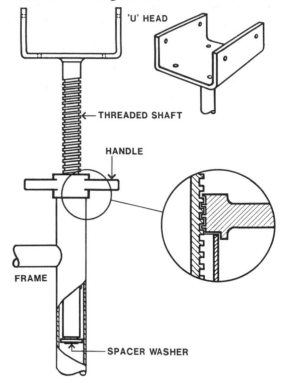

'U' HEAD

THREADED SHAFT

HANDLE

FRAME

SPACER WASHER

Figure 6.52 - TYPICAL 'U' HEADED SCREWJACK

Concentricity of the shaft in the tubular frame leg is essential to achieve its maximum load carrying capacity. In most cases the internal diameter of the frame leg is greater than the outside diameter of the shaft. To ensure concentricity a spacer washer is permanently fixed to the base of the shaft. This is shown in the cutaway view. The washer is sized to give a neat sliding fit on the inside of the frame leg.

The base of the frame requires a baseplate to transmit its load to the formwork footings. Even if the formwork is erected on a concrete slab a baseplate is needed. Bearing the end of the tube directly on the slab will result in damage to the end of the tube and unsightly ring marks on the slab.

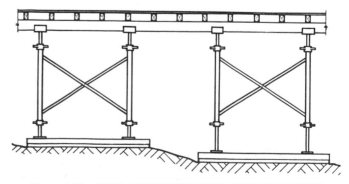

Figure 6.53 - BASE ADJUSTMENT FOR UNEVEN GROUND

The simplest, and a non-adjustable, base is the 150 mm square footplate with a short tubular spigot that inserts into the lower end of the frame leg. These were previously described with standard steel scaffold tubes as props. More commonly, flat headed screw jacks are used. As noted above these are general similar to the 'U' heads. They have the advantage of adding to the height of the assembly and, where the ground levels are variable, their extensions can be adjusted to suit. (Figure 6.53)

The importance of the connectors between successive stories of frames has been noted previously. There are many types of connectors, Figure 6.54 illustrates a typical one, showing the components on the left and the completed joint on the right.

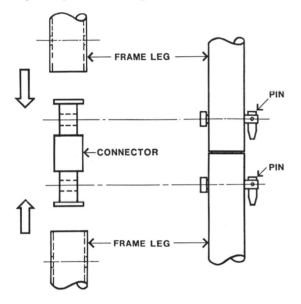

Figure 6.54 - TYPICAL FRAME CONNECTOR

The functions of connectors are to accurately retain the alignment of the upper tubular leg over the lower and to provide the necessary tensile connection between them. For this latter vital purpose the connectors are pinned to the frame legs. These pins can be simple bolts and nuts or the headed pins with snap-over ends as shown.

Horizontal Load Bracing for Frames

The lightweight erection and anti-buckling bracing that is installed on both sides of every pair of frames has already been described. This bracing is not adequate to provide resistance to the wind and construction forces which may act along the line of this bracing. For forces in this direction the formwork must be connected to the permanent structure (the concrete work which has been built) or have extra bracing added.

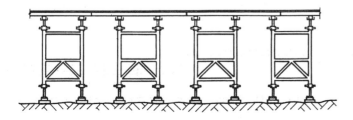

Figure 6.55 - END ELEVATION OF FRAME SETS

For forces acting along a line parallel to the frames, the internal bracing of the frames can often be adequate for a single storey frame assembly that is at least four rows of frames wide. (Figure 6.55)

It must be realised, however, that wind loads on formwork vary greatly with height and exposure of the formwork, (Figure 1.26) and every case should be checked for adequacy by engineering calculations.

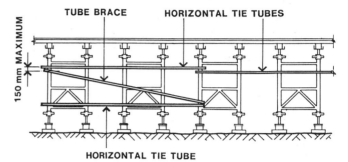

Figure 6.56 - EXTRA BRACING PARALLEL TO FRAMES

If additional bracing is needed for this direction of the single storey frame form this is usually framed from scaffold tubes as shown in Figure 6.56.

In the direction parallel to the erection braces, some extra bracing is needed if the form is not connected to the structure. Scaffold tubes are commonly used and are framed as shown in Figure 6.57. All these braces and ties must be coupled to the frame legs at all passings.

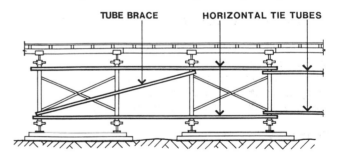

Figure 6.57 - BRACES BESIDE ERECTION BRACING

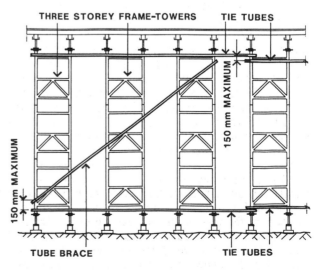

Figure 6.58 - BRACING A THREE STOREY FRAME SET

As stated previously, these frames can be erected many stories high. In general, such tall assemblies must have horizontal ties every third frame storey. If these ties are not linked into a suitable existing structure or building, which is adequate for the horizontal forces, diagonal braces must be added. Figure 6.58 shows an elevation of a typical case for forces parallel to the frames.

For forces parallel to the erection braces between frames the requirements are similar. An elevation of this is shown in Figure 6.59.

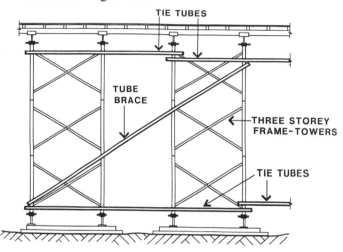

Figure 6.59 - SIDE BRACING TO THREE STORIES

As stated above, the horizontal tie tubes and raking brace tubes must be coupled to the frame legs at all passings. Splices in the long runs of horizontal tie tubes must not be made with end-to-end couplers. These have a very unreliable tensile capacity. Splices must be lapped, usually 500 mm, with couplers at each end and always located at a frame leg. Where possible, raking braces should be made of a single length of tube.

This has not been an exhaustive discussion of the details of supplementary bracing of tall frame assemblies. Reference should be made to the manufacturer's literature and advisory service for bracing requirements.

Load Application to the Frames.

At the commencement of this section on support systems, an example was given to show how the direction of the framing members can control the effect of the eccentric loads. This topic was illustrated with Figures 6.27, 6.28 and 6.29 and will be discussed in greater detail here. The matter is especially relevant to the safe and efficient use of frame systems. Knowledge of the reaction of the frame systems to the directions of the bearers at the top, and the soleplates at the bottom, can aid the formworker in achieving the safest and strongest structure.

If both the bearers and the soleplates are spanned from one frame to the other, that is, at rightangles to the plane of the frame, a situation like that portrayed in Figure 6.60 can occur. This front elevation of the frame gives a worst-case scenario for eccentricities in this direction.

At both the top and bottom the eccentric moments are all acting in the same direction. The resultant accumulated bending effect carries through the frame. The amount of sideways deflection will depend on the efficiency of the internal bracing of the frame and the stiffness of its frame

legs. For the case shown, where its internal bracing is not very effective, the deflection will be large. Nevertheless, any internal bracing, however arranged, acts with the frame legs and adds to the overall resistance to eccentric load effects.

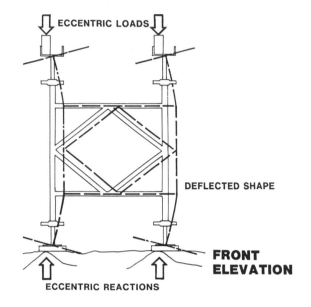

Figure 6.60 - ECCENTRICITIES ALONG THE FRAME

If, however, the bearers and the soleplates are run parallel to the plane of the frame, its worst case scenario of eccentricities is shown in the part side elevation of Figure 6.61. The eccentricities at both the top and bottom for all four legs of the pair of frames are all acting in the same direction. Both frames will deflect sideways in the same direction. Laboratory load tests on this combination of frames and eccentricities confirm that the lightweight erection bracing between the frames offers no restraint to this deflection.

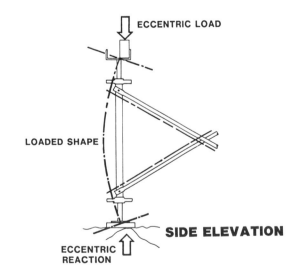

Figure 6.61 - ECCENTRICITIES ACROSS THE FRAME

Further, the internal bracing between the frame legs, being at rightangles to the direction of the eccentric moments, contributes virtually nothing to the frame strength. **The frame is at its weakest, when loaded in** this direction. This is not to say that it has no strength, but just that its load carrying capacity is far less than that which is achieved when the frame is loaded as shown in Figure 6.60. The ratio of the weakest to the greatest load capacity varies from frame type to type but tests often show this ratio to be less than one half.

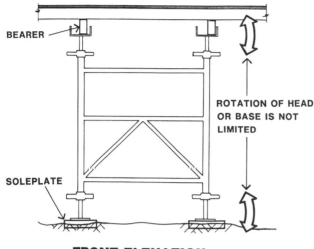

FRONT ELEVATION

Figure 6.62 - NO LIMIT TO TOP & BOTTOM ROTATION

Figure 6.62 again shows the recommended arrangement with the eccentricities acting in the strongest direction of the frame. Nevertheless, the rotation of the 'U' head at the top of the screwjack supporting the bearer can not be regarded as having any limition. The only restraint to this unlimited rotation is the stiffness of the slender shaft of the screwjack. Similarly, if the soleplate bears on a soft foundation there is very little effective limitation on the rotation of the screwjack baseplate. Naturally, the further the screwjacks are extended from the frame legs, the less the restraints are.

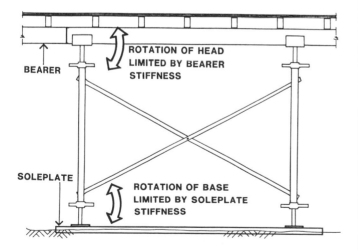

SIDE ELEVATION

Figure 6.63 - TOP & BOTTOM ROTATION IS LIMITED

But it must not be thought that this bearer and soleplate framing arrangement completely eliminates all the eccentric effects in the weakest direction of rightangles

to the plane of the frames. Figure 6.63 shows the side elevation of the recommended arrangement.

At the top of the frame some rotation (and eccentric) action will occur due to the normal deflection of the bearer under load. This deflection is designed to be small so that the slab soffit will be within tolerances. A bearer deflection of 3 mm would rotate the 'U' head an almost insignificant amount.

Of greater importance is the load response of the soleplate. If the ground is soft, and the soleplate flexible, deformation can occur as shown previously in Figure 6.20. The resulting baseplate rotation can severely reduce the frame's load capacity. With the soleplate appropriately stiffened, as was shown in Figure 6.22, its deflection is very small and the baseplate rotation and resultant eccentric load effect can become insignificant.

From all of this discussion it will be obvious that when frame systems are erected on concrete slabs, their load capacity is greater than when they are founded on soil, regardless of the soleplate stiffness. Even if the concrete surface is uneven as was shown in Figure 6.17 there can only be a small and limited baseplate rotation. With the eccentric effects thus almost limited at the frame bases, the load capacity improves. Nevertheless, the bearer directions should still be as recommended.

Frame Systems Erection Procedures.

Because of the inherently stable nature of each braced pair of frame sets, extra bracing precautions only have to be taken during erection if the assembly is three or more frame storeys high.

After the completion of the soleplates or other temporary footing work, the pairs of frames are erected. Using two-man crews, each frame is stood up on its base screwjacks and the lightweight erection braces added to connect it to the other frame.

For single storey assemblies the top components, 'U' heads or flat headed screwjacks, are added. Then some scaffold planks are often installed on the top or middle horizontal rail of the frames to provide a safe work area for the formworkers installing the bearers and joists. After these are in place, bundles of plywood can then be hoisted onto the joists and their placing carried out as described for prop formwork. After completion of this stage, the deck can be adjusted to level and any needed supplemetary tube ties and braces added.

A Multi-storey Formwork Assembly

Where multi-storey frame assemblies are being erected, the supplementary ties and braces must be installed at each three frame storey interval or closer. For example, for a four frame storey assembly, the ties and braces would be installed after the completion of the first two storeys and then at four. For tall narrow assemblies, external guys and extra supplementary tube ties and braces may be needed. The key is, safety at all times. (Refer to Figure 1.06)

Modular Systems

There are a number of modular support systems available from formwork manufacturers. This description of modular systems will be focussed on the essential features that most of these systems share. When using a particular sytem, reference should always be made to the manufacturer's technical literature.

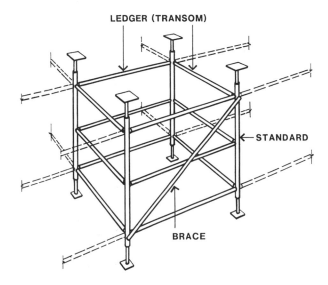

Figure 6.64 - TYPICAL MODULAR SUPPORT SYSTEM

Figure 6.64 diagramatically shows the general arrangement of one bay of a modular system. It is an assembly of individual standards and end fittings, horizontal tie members, and raking braces. Standards (tubular columns), which are available in a range of modular heights, are connected at the top, bottom and regular spacings between, by tie members known either as transoms or ledgers.

Use of the appropriate number of levels of ties is crucial to the safety of modular form systems. To leave out a level of ties with a resultant doubling of the buckling height will lead to the load capacity becoming one quarter of its expected value.

Similar to frame systems, 'U' head or flat headed screwjacks are inserted at the top of the standards and flat screwjacks at the bases. The other end fittings described for use with standard steel scaffold tube propping (base plates, bearer support fittings etc.) are also in general use with modular systems.

In general, transoms are members that not only control the spacing and buckling length of the standards, but are also designed to carry scaffold planks to provide work access. Commonly these are made from two steel angles welded together to give an inverted Tee section that provides a seating for the scaffold planks. Figure 6.65

shows a cross-section through a transom with the planks that it supports.

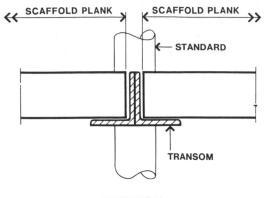

SECTION

Figure 6.65 - TRANSOM SUPPORTING SCAFFOLD PLANKS

Tubular ties, on the other hand, usually only control the spacing and buckling lengths of the standards. To enable the standards to erected on a range of different size grids, the ties (transoms or ledgers) are available in lengths ranging from 0.9 m to 2.4 m and the ties from 1.8 m to 3.0 m.

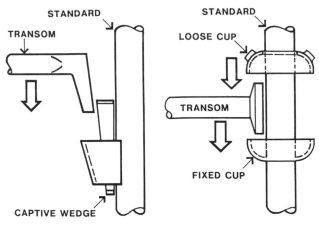

Figure 6.66 - TWO TYPES OF TRANSOM FIXINGS

The transoms are connected to the modular positions on the standards with quick acting fixings. Figure 6.66 shows two typical types of fixings. The left-hand sketch shows the transom provided with a tapered spigot that inserts into a loop on the standard, and is retained by the tightening of a captive wedge. To release the transom, the wedge is hammered upwards and the transom simultaneously lifted out. These fixings provide quite rigid joints and this feature is useful during the erection procedure.

The right-hand sketch shows a method that fixes or unlocks the ends of up to four transoms in the one operation. The transoms have 'Tee' headed ends which are hooked into a fixed cup on the standard. A loose inverted cup on the standard is then dropped over the tops of the 'Tee' ends to the ties. The upper cup is provided with lugs so that it can be rotated with a hammer to tighten onto all the transom ends.

When tall formwork is needed, the standards can be assembled from a number of shorter modular standards joined concentrically with connectors inserted in their mating ends. Like the connectors used with frames, pins must be inserted to give the junction tensile capacity.

The raking braces are usually scaffold tube size and connect to the same points that the transoms use. Braces, which must act in two directions at rightangles, are to be installed for the full height of the assembly.

Braces on the outside of the outer row of standards can be connected to coincide with the levels of the transoms. The concurrency of action of standards, transoms and braces makes them structurally efficient. However, raking braces within the form assembly cannot usually achieve this concurrency of action.

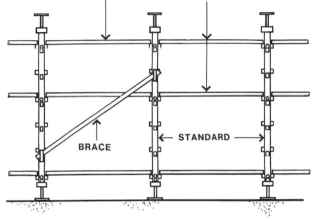

TRANSOMS IN BOTH DIRECTIONS

BRACE
STANDARD

Figure 6.67 - SECTION THROUGH MODULAR SYSTEM

As shown in Figure 6.67, to avoid conflict with the transoms, these internal braces join the standard at a level one module below or above a transom line. To compensate for the resultant reduction in structural efficiency, extra braces are often needed. The bracing requirements for each system are to be found in the manufacturer's technical literature.

Modular Systems Erection Procedures.

Similar to the procedures for props, the erection of modular frames starts with a framed and braced core. For speed of assembly the lines of standards can be erected out from this core by installing only the top transoms to the standards. (Figure 6.68)

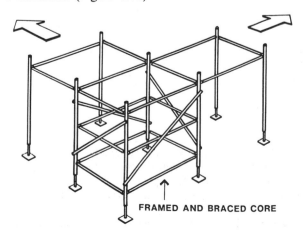

FRAMED AND BRACED CORE

Figure 6.68 - ERECTION STARTING AT A BRACED CORE

A second erection crew or crews can then follow this up, installing the bottom and intermediate ties and transoms together with more braces. The needed scaffolding planks can also be fitted to the transoms to provided elevated work areas. For multi-storey formwork assemblies, the lower framing storeys should each be completed and braced before proceding to the next.

After tying and bracing, the standards can have the top fittings installed followed by the bearers, joists and plywood.

Combination of Frames and Props

In Chapter 1 a formwork system where the supports were a combination of alternate rows of props and frames was shown in Figure 1.20. This illustrated the situation where strict precautions were needed in the placing of heavy loads on the partially completed structure. Figure 6.69 illustrates the sequence of erection.

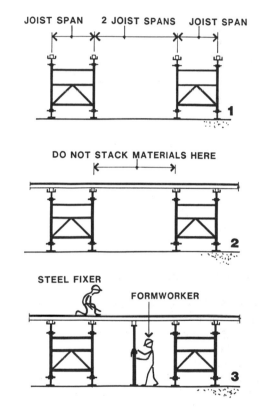

JOIST SPAN | 2 JOIST SPANS | JOIST SPAN

DO NOT STACK MATERIALS HERE

STEEL FIXER
FORMWORKER

Figure 6.69 - ERECTION OF COMBINED FRAMES & PROPS

The procedure starts with rows of frames which are located to have a space of twice a joist span between them. After the bearers are erected on the frames, single span joists are placed over the frames and double span joists between them. Bundles of plywood sheets can then be hoisted onto the frames; **but not out on the double span joists!**

After the plywood has been placed, other work such as edgeforms, penetrations and steel fixing can be done while the rows of props and their bearers are installed between each row of frames. Any needed supplementary bracing can then be installed.

In calculating the size of the joists, the double span situation will have to be considered for a combination of the dead load of the form, the live load of formworkers and an allowance for tools and materials. Obviously it would be uneconomic to design the joists, in this double span situation, for bundles of reinforcement and the like. The formwork drawings must contain clear warnings about this, and instructions to place these loads directly over the frames.

The principal advantage of this method is speed of erection and the earlier commencement of steel fixing and other following trades.

GENERAL CONSTRUCTION DETAILS

Junctions at Concrete Walls.

Figure 6.70 shows a section through a typical junction of a soffit form and a wall. In this case a scaffold tube prop with a 'U' head is shown.

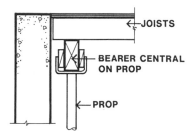

Figure 6.70 - PROPPING AT A WALL FACE

An examination of this detail will show that there is nothing to prevent the form moving away from the wall face. General restraint would be provided by the normal bracing to the overall formwork structure, but a more positive local holding method is needed to stop a small gap opening. One simple method would be a series of raking telescopic props. (Figure 6.71)

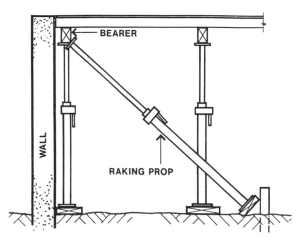

Figure 6.71 - HOLDING THE FORM AGAINST THE WALL

Alternatively, if the wall is reinforced concrete, the wall tie-rod locations for the wall forms can be used to hold the formwork against the face of the concrete wall. Details for both bearers and joists parallel to the wall are shown in Figure 6.72. Horizontal movement is prevented by the clamping action at the wall face. However, the vertical loads on the formwork are still carried on props or frame assemblies.

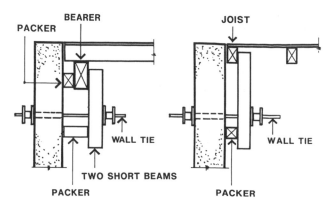

Figure 6.72 - FORMWORK CLAMPED TO FACE OF WALL

Where these details are inconvenient, the full support of the form on the wall can be considered. Figure 6.73 shows two details. In the left hand detail, the wall ties are used this time for both support, and retention to the wall face. The first packer gives the bearer a seating, the second extends up its side to provide retention.

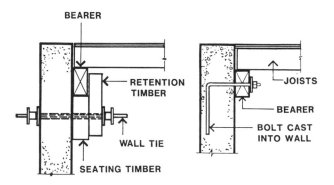

Figure 6.73 - CARRYING THE FORM ON THE WALL

In the right hand detail, bolts have been cast into the wall to bolt the bearer directly to its face.

Edge Forms

The depth of the edge of suspended structures that has to be formed can vary greatly. For slabs of 100 to 200 mm thickness, the edge form can simply be a solid timber section, usually 50 mm thick. As Figure 6.74 shows it can be fixed in place by skew nailing to the formface and should be braced at approximately 1.5 m centres. These braces are often off-cuts of formply.

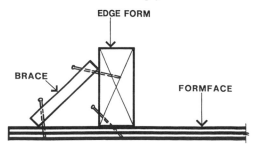

Figure 6.74 - A TYPICAL SHALLOW EDGE FORM

Deeper edge forms, 300 mm or more, usually need to be framed and sheeted with plywood. Figure 6.75 shows a section of a deeper edge form with top and bottom plates and studs, usually at 450 mm spacings.

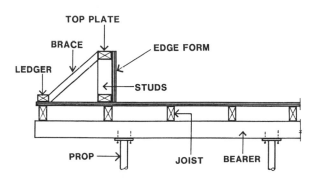

Figure 6.75 - A TYPICAL DEEP EDGE FORM

With the deeper forms, the forces on them from the fluid concrete are commensurately greater. To prevent the outer sheets of soffit form plywood sliding under the effects of the forces on the edge form, extra attention to the nailing is needed. Also, the braces require better fixings and a ledger is provided for this.

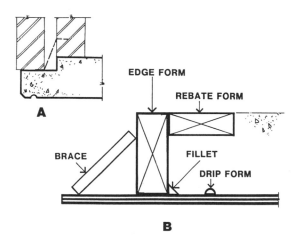

Figure 6.76 - REBATED EDGE TO CONCRETE SLAB

The edges of structure often require a rebate to be formed in the slab. Detail 'A' of Figure 6.76 shows this type of construction requirement for a brick wall. Detail 'B' shows the rebate formed with a continuous solid timber section. Note the arris form (fillet) to give a chamfered edge to the slab, and the form for the drip groove on the soffit of the slab.

Construction Joints

Construction joints are planned interruptions to continuous concrete structures. Reinforcement is usually continuous through the joint. For economy of construction the formwork at the construction joint needs to be removed early so that the next stage of the work can procede. As this stripping usually occurs while the concrete is young and weak, the forms must be designed for easy and gentle removal.

Looking at a difficult system first. Figure 6.77 shows a keyed construction join formed from three continuous timber strips. The reinforcement passes between these strips. The lower strip is equal in height to the

reinforcement cover as is the top one. Short braces are provided to maintain the verticality of the form. When the concrete is placed some concrete will intrude into the gaps between the timbers.

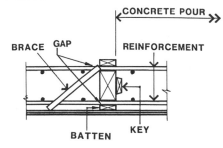

Figure 6.77 - SLAB CONSTRUCTION JOINT - 1

At stripping, the top timber strip is easily removed and the concrete between the two top timbers is readily chipped away. However, getting the middle timber out usually involves levering it sideways and, as a result, the top reinforcement is often lifted slightly and the top concrete is spalled. The removal of the bottom strip is even more difficult.

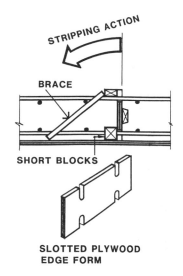

Figure 6.78 - SLAB CONSTRUCTION JOINT - 2

A construction joint form shown in Figure 6.78 is much easier to remove. The formface comprises an optional key form fixed to a plywood face. The plywood is slotted as shown at the reinforcement positions. This form is held to line by a continuous timber along the top, and a line of short timber blocks on the soffit form, between the reinforcement. Short braces to the top timber maintain the top line and plumb of the face.

When stripping the construction joint form, the braces, top timber and timber blocks are readily removed. The plywood face can then be tilted over and removed from between the layers of reinforcement. With care, no damage to the concrete occurs.

Following this, when the second stage of the concrete placement occurs up against the construction joint, grout loss may occur at the construction joint. If a construction joint is located near the middle of the formwork framing spans, a further deflection will occur when the second stage is poured. The previously placed section of the slab, having developed concrete strength, will not deflect with the formface. Grout will intrude under the earlier pour as indicated in Figure 6.79. To prevent this deflection, an extra line bearers and propping may be needed along the line of the construction joint.

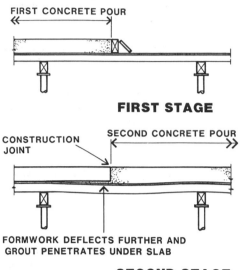

FIRST STAGE

SECOND STAGE

Figure 6.79 - SECONDARY DEFLECTION AT THE JOINT

A further problem can occur with the stacking of materials alongside a construction joint. The three stages of this are illustrated diagramatically in Figure 6.80.

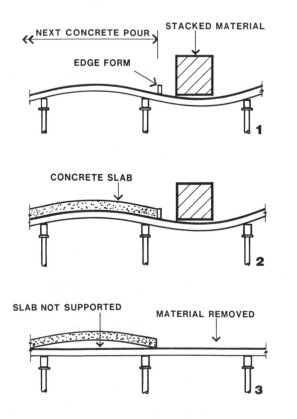

Figure 6.80 - MATERIAL LOADS NEAR A JOINT

At first, the material has been stacked alongside the construction joint before the first pour; the forms deflect

downwards under this load and hog upwards in the adjacent span. When the concrete is placed it tends to slightly offset this hogging but it still develops its strength on a form which is deflected upwards.

When the material load is removed so that the next pour can be got ready, the formwork reverts to a generally level shape. This leaves the slab without support. Depending on its age, cracking of the slab may result.

Top Steps in Slabs

It is often necessary to provide an edge form for a step in the top of a concrete slab. If the step is near the edge of the slab and the soffit formwork extends well beyond the edge form, the detail given in Figure 6.81 can be used.

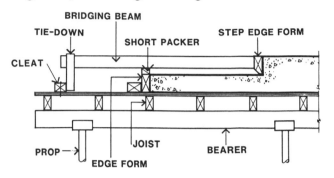

Figure 6.81 - TOP STEP FORM NEAR THE SLAB EDGE

In this method the step form is cantilevered over the edge form. Short packers lift the levels of the bridging beams above the concrete to provide trowelling access. A tie-down and cleat are fixed to the formface and each bridging beam to hold the step form to level.

Where the soffit form does not extend enough to give a tie-down, the method shown in Figure 6.82 can be used.

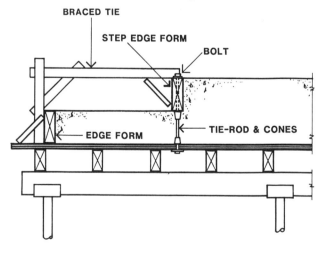

Figure 6.82 - TOP STEP FORM ON TIE-RODS AND CONES

The level of the step edge form is maintained by supporting it on wall formwork hardware: tie-rods, cones and He-bolts. The line of the step form is controlled by a braced tie connected to the outer edge form.

Where the step is located a distance from the edge of the pour it can still be supported by wall form hardware, but needs a method like that shown in Figure 6.83 to maintain its line.

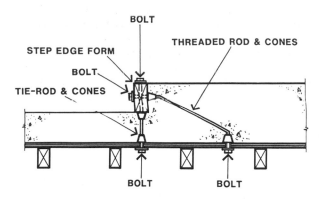

Figure 6.83 - STEP FORM AT INTERIOR OF SLAB - 1

A threaded rod is bent to shape and fitted with cones. The cones are bolted to the step form and the plywood soffit. A simplification of this is shown in Figure 6.84.

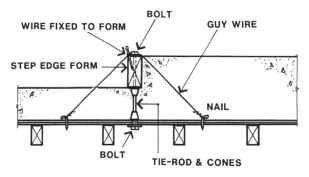

Figure 6.84 - STEP FORM AT INTERIOR OF SLAB - 2

The step form is shown braced with tie wires (usually 1.6 mm diameter) which pass through hooks on the soffit form. These hooks are usually 60 mm nails bent over. The tie wires are fastened at the top of the edge form. These braces are not very strong and are usually needed at 1 m spacings.

After the concrete has achieved its initial set and surface trowelling has commenced, the wires can be cut at the top to a short distance below the slab surface. When the soffit forms are stripped, the projecting nail ends are cut flush with the concrete soffit. One minor, but less satisfactory, variation on this is to tie the form to the reinforcement.

Slab Penetrations

Forming rectangular penetrations through the slab is merely an application of edge form techniques.

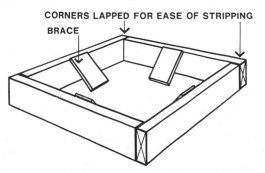

Figure 6.85 - RECTANGULAR PENETRATION FORM

Figure 6.85 shows an example. Note the sequence of lapping at the corners, this aids the stripping in a way similar to that shown in Figure 4.102 for wall penetrations.

Where the penetration forms define the inside of a manhole cover, nails are driven into these edge form to maintain the line of the top of the cast iron frame at slab level. Figure 6.86 shows a section through one of the edges of the penetration forms.

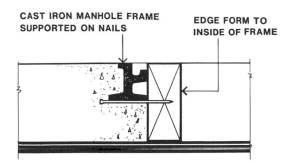

Figure 6.86 - PENETRATION FOR MANHOLE FRAME

Smaller penetrations, rectangular or otherwise can be made from blocks of rigid foam plastic. Being very light they need to be firmly anchored. Sliding can be prevented by forcing the foam block down over four or more large nails driven into the formface. Uplift and overturning can be inhibited by tying wire guys over the block and down to the bottom slab reinforcement. (Figure 6.87)

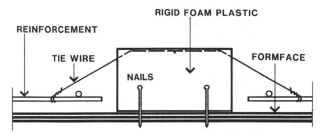

Figure 6.87 - FOAM PLASTIC PENETRATION FORM

Circular penetrations are readily formed in sheet metal. Details of this method are shown in Figure 6.88

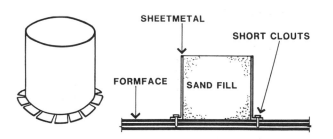

Figure 6.88 - CIRCULAR PENETRATION FORM

The flange of the penetration form is fixed to the form face with short clouts. These will give a limited amount of fixing so that the stripping of the soffit form will not be inhibited. To prevent the form being filled with concrete and limit the amount that it is pushed out of shape during the pour, it is recommended that the form be filled with sand.

Small pipe penetrations can be made with plastic pipes. Figure 6.89 shows one method of installation.

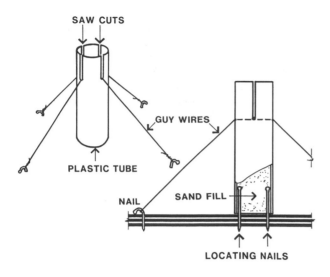

Figure 6.89 - SMALL PIPE PENETRATION

The base of the pipe can be held in position with a group of nails driven into the formface. The pipe is held vertical with tie wire guys to nails in the formface. Note that the wires are set in sawcuts to prevent them being dislodged. They are also placed below the top of slab so that they will not interfere with the trowelling of the concrete surface. Similar to the other case, the pipe should be filled with sand.

Minor Setdowns in Soffit Forms

Simple techniques are available for the forming of minor steps in soffit forms. Minor setdowns can be broadly defined as being 200 mm or less. Figure 6.90 shows sectional views of three examples of setdowns in the concrete soffit.

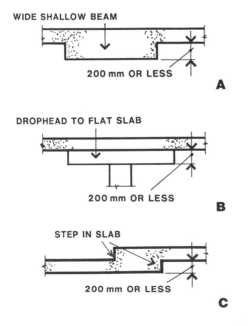

Figure 6.90 - EXAMPLES OF MINOR SETDOWNS

Example 'A' is a wide shallow beam, 'B' is a drophead panel to the column under a slab and 'C' shows a simple change in slab level.

The technique simply involves lapping the higher level soffit form onto the lower. Figure 6.91 shows the soffit step parallel to the bearers.

To achieve the required step a packer timber is stripped to the appropriate height, and the joists on the upper level are seated on this packer. The open edge can be closed off with a plywood strip.

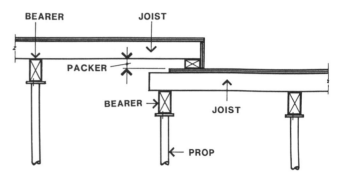

Figure 6.91 - STEP PARALLEL TO BEARERS

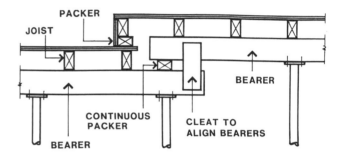

Figure 6.92 - STEP PARALLEL TO JOISTS

If the soffit step is parallel to the joists, the detail given in Figure 6.92 is suitable. This shows the bearer from the upper level supported on the lower bearer through a continuous packer member. The packer is continuous to cater for any minor eccentricities of loading between the two bearers. To maintain alignment of the upper to the lower bearer, they should be securely cleated together with the plywood off-cuts.

The above two details are applicable to the edges of a slab drophead panel at columns.

Sloping Formwork

When concrete is placed on sloping forms or the formwork supports are founded on a sloping surface there are resolved forces acting along the lines of the slopes that must be countered by braces and anchorages. Further, the seating of the bearers on the cap plates or 'U' heads of supports must still be constructed to be as concentric as possible.

Some proprietary brands of equipment, as shown in Figure 6.93, have a tilting 'U' head, sometimes called a 'rocking forkhead'. However, the hinging action of these is usually limited to 10 degrees of slope.

To prevent the joists sliding or rolling over they should be skew nailed to the bearers and blocked as shown. The frequency of blocking needed will depend on the slope and the loadings.

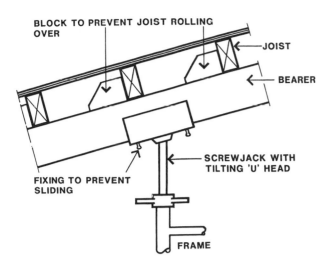

Figure 6.93 - SLOPING FORM WITH TILTING 'U' HEAD

If the articulated 'U' head is not available, or the slope is too steep for it, then concentric load action on the support system can be achieved by details like that shown in Figure 6.94.

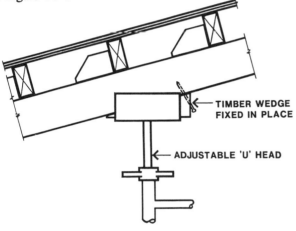

Figure 6.94 - SLOPING FORM WITH WEDGE ON 'U' HEAD

A wedge between the bearer and the 'U' head, securely nailed in place to prevent movement, achieves alignment between the sloping bearer and the level 'U' head. Concentricity in the other direction is obtained by rotating the 'U' head and securing this with a nail into the side of the bearer.

Within the support framework the formwork designer will have to pay particular attention to providing adequate bracing to prevent the supports from moving along the line of the slope, under the influence of the induced sloping forces. A tube tie along the slope, coupled to the shafts of all the screwjacks just below the sloping bearers, is often needed to cater for the sloping forces.

At the footing level two major concerns must be addressed: the stability of the soil slope and the lateral stability of the baseplates to the supports.

The load bearing characteristics of the slope must be carefully assessed for stability. Usually a geotechnical investigation is needed. Soft soils can either fail in bearing or, as shown in the left hand sketch of Figure 6.95, a segment can rotate in what is known as a 'slip-circle' failure.

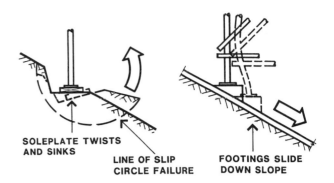

Figure 6.95 - TWO FAILURES ON A SLOPING BANK

If adequate anchorage is not provided, the base of the support may slide down the slope. (Figure 6.95, right hand sketch) The resultant buckling failure of the support can trigger a progressive collapse.

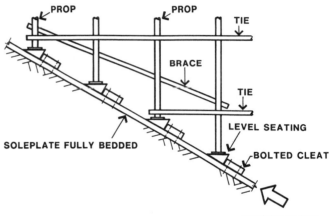

Figure 6.96 - SLOPING FOOTING FOR FORMWORK

Assuming that the adequacy of the soil has been proved, the elevation shown in Figure 6.96 gives general details of a suitable footing arrangement. The sole plate must firstly be adequate in size and fully bedded to the ground. Cleats bolted to the soleplates carry the thrust from the seating blocks. At the bottom of the soleplate, an adequate thrust block must be provided to cater for the accumulated forces from the formwork supports.

Supporting Soffit Forms on Steel Beams

Flat concrete slabs are often supported on structural steel beams. Figure 6.97 shows the cross sections of two common types of this.

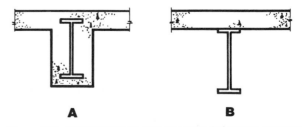

Figure 6.97 - CONCRETE SLABS SEATED ON STEEL BEAMS

Detail 'A' shows the concrete encasing the steel beam; this is usually done for fire protection. The formwork requirements for this case are covered in Chapter 7 and shown in Figure 7.58. Where fire protection is not required the slabs are seated on the beams as shown in detail 'B'.

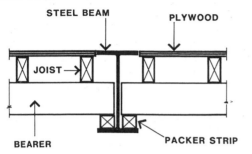

Figure 6.98 - SOFFIT FORM SEATED ON LOWER FLANGE

Provided the steel beams are deep enough, the bearers can often be carried on the lower flange of the beams as shown in Figure 6.98. It is necessary to get the approval of the design engineer for the steelwork, before placing the bearer load on the bottom flange of the beam. Note that the bearers are packed to bring the level of the top of the plywood (slab soffit) to that of the top of the beam. In the details that follow, the other common practice of aligning the slab level with the underside of the top flange will be shown.

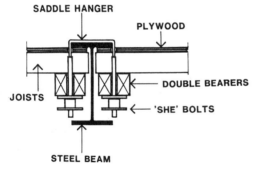

Figure 6.99 - SOFFIT FORM SUSPENDED ON HANGERS

Figure 6.99 shows the cross section on the formwork suspended from the steel beam on saddle hangers. These are bent tie rods with threaded ends. She-bolts (refer to Chapter 4, Figure 4.23) connect to these saddle hangers and support pairs of bearers. The whole assembly is held firmly in place by tightening the plywood up against the underside of the beam flange.

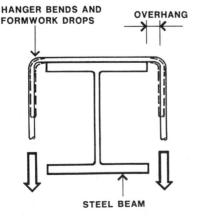

Figure 6.100 - HANGERS PROJECTING BEYOND BEAMS

It is essential that the vertical legs of the saddle hanger fit tightly up against the edge of the flange. Figure 6.100 illustrates the inevitable failure that occurs when the saddle hanger overhangs beyond the edges of the steel beam.

If the available sizes of saddle hangers are wider than the beam flange then they should be placed in a skew position on the top flange of the beam. This eliminates any gap between the flange edges and the hanger legs. The plan view of the beam given in Figure 6.101 illustrates this.

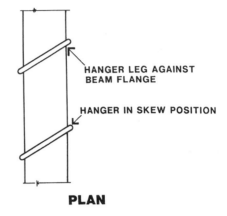

PLAN

Figure 6.101 - HANGERS TIGHTLY AGAINST BEAM SIDES

Slabs supported on steel beams commonly project beyond the outer row of beams. The formwork for these areas must also cantilever beyond the steel beams. In Figure 6.102 a method suitable for short overhangs is shown.

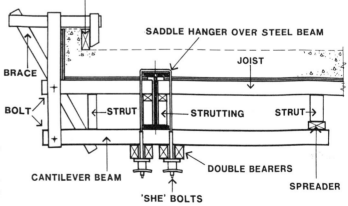

Figure 6.102 - CANTILEVERING SOFFIT FORMWORK - 1

The cantilever beams, at the same spacings as the timber soffit form joists are clamped to the underside of the outer beam. The form joists are supported at this steel beam on timber plates strutted up from the beam's lower flange.

The uplift forces on the inboard end of the cantilever beams act through a continuous spreader timber onto struts up to each joist. The spreader timber is essential to counter any twisting effect that could occur from a small misalignment of the cantilever beams, struts and soffit form joists. All junctions between these timber members must be securely nailed. The upthrust forces from the inboard end of the cantilever beams will tend to bend the soffit form joists upward, and push them against the underside of the top beam flange.

As shown, the forms to the outer slab and its edge are strutted up from the outboard end of the cantilever beam. As noted above, this detail is only suitable for short overhangs due to obvious safety limitations on upthrust forces on the joists. A method that can cater for a greater overhang is shown in cross section in Figure 6.103.

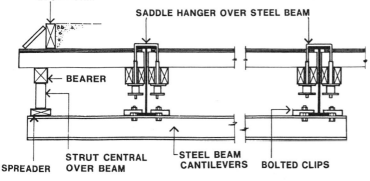

Figure 6.103 - CANTILEVERING SOFFIT FORMWORK - 2

For the support of the cantilever form section, steel beams are clipped to the underside of the two outer rows of steel beams. Their spacing does not relate to the joist spacing but is determined by the permitted spans of the bearer at the outer edge of the form.

The steel beam cantilevers are suspended from the main beams by bolted clips. Similar to the previous method for short overhangs, the struts on the beam are seated on a spreader timber to cater for any minor misalignments of the struts from the bearer and the centre-line of the steel beam cantilevers. As before, all junctions must be securely nailed.

With both these methods for forming overhanging slabs, it is essential that the structural sizes and connection methods be checked by an engineer competent in structural design.

A Tolerance Problem

There are many cases like those shown in Figures 6.98 and 6.99, where the formwork has to fit between two fixed lines: between walls or between beam faces. Two more cases of this are shown in the sectional views of Figure 6.104.

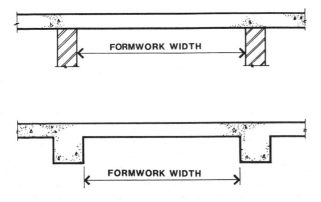

Figure 6.104 - FORMWORK MUST FIT BETWEEN FACES

Only rarely can this width to be formed be made up of the standard sizes of plywood sheets. Usually one row of sheets will have to be cut down to a width that makes up the difference between the width to be formed and a multiple of standard plywood sheet widths.

Often this 'make-up' row of plywood sheets can be advantageously placed in the middle of the work as a stripping band. (Figure 6.105) The removal of this strip first, makes stripping of the other sheets easier. Stripping of soffit forms is covered later in this Chapter.

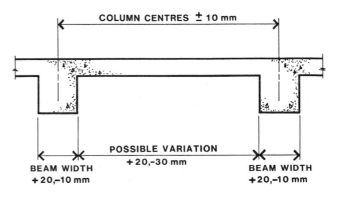

Figure 6.105 - FORM MAKE-UP AND STRIPPING BAND

For a single-use construction of formwork the cutting of a 'make-up' strip is normal practice. However, if this formwork construction has to be repeated a number of times, such as the successive floors of a high-rise building, then two problems can intrude.

Firstly, although each formwork area is intended to be the same width as the others, there can be deviations in dimension and position of parts of the structure. This is not to say that the work is inaccurate and unacceptable. These deviations may be well within tolerances. Figure 6.106 shows an example of a soffit form between two beam faces. As tolerances are usually defined for each case by the design engineer for the structure, the values given should not be taken as typical of all work.

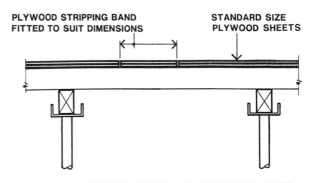

Figure 6.106 - EXAMPLE TOLERANCES FOR CONCRETE

Assuming the beams to be symmetrically placed on the columns in each case, the soffit formwork width can be 20 mm wider or 30 mm narrower than the specified dimensions. If the beams are not always concentric on the columns then the range of widths could be even greater. Obviously many cases of cutting the 'make-up' strip narrower or cutting new material for a wider gap could occur.

Secondly, at each pour, a small amount of grout often penetrates the small gaps that will be found between adjacent plywood sheets. In the first instance these gaps will be a natural result of the permitted manufacturing tolerances in the dimensions of plywood sheets. When the plywood is stripped some of this hardened grout sticks to the edge of the sheets. Effectively they get larger; they 'grow'. At the next use, the gaps that the irregular edges of the sheets create, are in turn filled with grout.

The net result of all this is that, even if the required formwork width were to remain exactly constant, there would usually need to be some trimming of the 'make-up' strip.

On many formwork jobs this is exactly what happens; the strip is recut or replaced as needed. However, this can be avoided. Figure 6.107 shows a cross section through detail that provides a 'tolerance gap' that can give the necessary variations in formwork width without recourse to wasteful recutting.

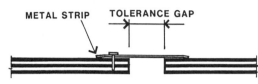

Figure 6.107 - FORMING A TOLERANCE GAP

The plywood widths can be arranged to leave one or more tolerance gaps of approximately 25 mm. Each of these is bridged by a metal strip, usually 0.8 mm thick, and fixed with short clouts. These fixings can be quite widely spaced as they are only to stop it being dislodged during the concrete placement. The only impediment to the use of this technique would be extremely stringent tolerances on face steps in the concrete soffit. This tolerance gap can also be incorporated in the edge of the beam forms. This is discussed in Chapter 7.

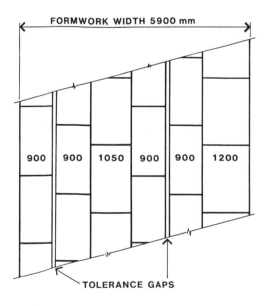

Figure 6.108 - PLAN OF A PLYWOOD SHEET LAYOUT

Figure 6.108 shows a part plan of an example of a form of 5900 mm width. To give adequate variations in width, and cater for permitted dimensional deviations, two 25 mm tolerance gaps are provided. The remainder is provided by four 900 mm wide sheets, two 1200 sheets

and one 1200 mm sheet cut to 1050 mm.

These tolerance gaps serve the further purpose of alleviating the compression on the plywood that comes from concrete shrinkage and the grout penetration between the plywood sheets. As a result they facilitate easier stripping.

FORMWORK STRIPPING

Formwork stripping, also called striking, is the process of removing the formwork. The earliest time when this can be done is when the concrete has developed sufficient strength for the reinforced concrete structural elements to be able to carry themselves, and any other superimposed loads, over the intended spans. These spans may be the full spans of the final structure or something less due to the installation of temporary supports such as props.

The act of stripping transfers part or all of the loads to the concrete structure. The strength of the concrete and the manner of that transfer can affect the deflection of the structure. Because of this potential effect on the serviceability of the reinforced concrete structure, the conditions to be met for stripping must be stipulated by the design engineer for the structure.

In addition to the general rule that all stripping shall be done such that there is no thermal or physical shock, and that there shall be a gradual transfer of load, the relevant matters which should be specified in the project documentation are:

- the concrete age and minimum strength at the time of stripping, taking into account the effect of the ambient temperature on the concrete strength growth.
- the effectiveness with which the curing procedures are carried out.
- limits on the superimposed loads to be placed on the structure.
- the method and sequence of stripping.

To achieve the economies that come from early formwork removal and reuse, stripping methods are usually a two stage process. The first stage, involving relatively short spans between props, can usually be done at time well before the development of the expected concrete strength.

It will become obvious from the descriptions given below, that the positions of the props can sometimes be quite random, Even when they are in a line, they do not readily relate to the reinforcement pattern of the concrete structure. Therefore, to avoid any detrimental cracking patterns, the lengths of the spans between the first stage props are usually calculated on what is known as the 'uncracked section'. This is controlled solely by the tensile strength of the concrete.

The methods that use this two stage process can be described under three headings: **Undisturbed Supports, Backpropping** and **Reshores**.

Undisturbed Supports.

Figure 6.109 shows the principles of the sequence of the two stage process of undisturbed supports. Details of some proprietary formwork systems that use this sequence are given later in this chapter. In this method the cap plates of the props are part of the soffit formface. The formwork framing, carrying the remainder of the formface, is attached to the side of the props not far below the cap plate.

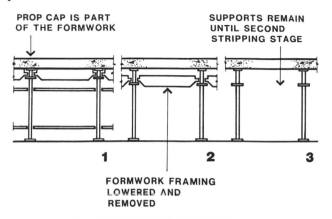

PROP CAP IS PART OF THE FORMWORK

SUPPORTS REMAIN UNTIL SECOND STRIPPING STAGE

1   2   3

FORMWORK FRAMING LOWERED AND REMOVED

Figure 6.109 - UNDISTURBED SUPPORTS

When the concrete has attained sufficient strength for the structure to be able to safely span the short distance between the supports, the first stage can commence.

The attachment of the framing to the formwork supports has a mechanism for lowering the framing. This mechanism always lowers the framing gradually at first so that the reinforced concrete receives the stresses gradually. In some proprietary brands there is a second stage of the mechanism that drops the framing and formface rapidly to be approximately 75 mm below the concrete soffit.

The concrete slabs now span between the supports and the formface, and framing can be removed for possible use elsewhere. The second stage of stripping the undisturbed supports can occur when the concrete strength gain is adequate, and the slab loading is within the specified limits.

The undisturbed formwork process does not have to involve sophisticated proprietary formwork systems. Conventional formwork can be simply left in place until the conditions for the second stage of the stripping are met.

Backpropping.

This is another method of maintaining some degree of continuous support of the slab, while getting the formface and its framing out at an early age for re-use elsewhere.

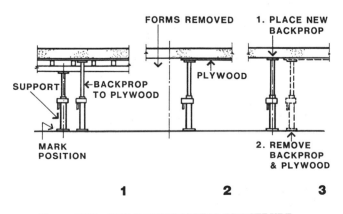

FORMS REMOVED

1. PLACE NEW BACKPROP

PLYWOOD

SUPPORT   ←BACKPROP TO PLYWOOD

MARK POSITION

2. REMOVE BACKPROP & PLYWOOD

1   2   3

Figure 6.110 - THE BACKPROPPING PROCEDURE

Figure 6.110 illustrates the sequence of backpropping. When the concrete strength is judged to be adequate for the first stage, backprops are placed in a suitable position alongside the formwork supports and tightened up against the plywood. The position of each of the formwork supports is marked on the floor for locating the second set of backprops. The formface, framing and formwork supports can then lowered and stripped out, leaving the concrete slab supported on the first set of backprops.

Above each of these backprops is one sheet of plywood. To enable removal of these sheets, a second set of backprops is placed, ideally in the marked position of the original formwork supports. The first set of backprops and their plywood sheets can then be removed.

Although there is some small amount of slab deflection, and load transfer, as each set of backprops shortens slightly when the load is transfered to them, continuous support is maintained for all practical purposes.

It is not always possible to place the second backprop in the position of the original formwork support. The plywood sheet is often also above this location. If so, the second backprop is then placed as near as possible to the original position and the resultant backprop layout can be quite random. To avoid this and eliminate one propping operation, backpropping strips can be used.

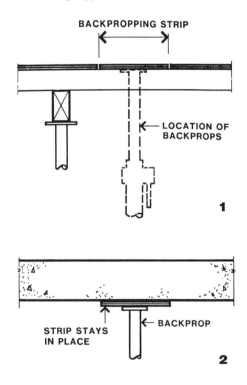

BACKPROPPING STRIP

←LOCATION OF BACKPROPS

1

STRIP STAYS IN PLACE

←BACKPROP

2

Figure 6.111 - THE USE OF BACKPROPPING STRIPS

Figure 6.111 shows a section through formwork where the formwork has been constructed with backpropping strips. These are narrow strips of plywood, say 300mm wide, located along the specified lines for the backprops. They can also serve as the 'make-up' strip discussed earlier. As shown, the backprops are placed along these strips where they remain until the second stage of stripping some time later.

By controlling the backprop locations in this way they can be positioned to relate to the reinforcement pattern. With careful consideration this can reduce stripping times.

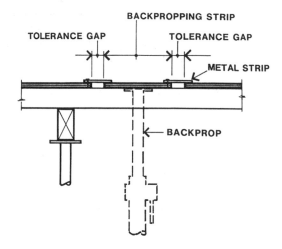

Figure 6.112 - BACKPROPPING STRIP & TOLERANCE GAPS

To cater for the tolerance problem discussed earlier, and ease the tightening effect that concrete shrinkage and grout penetration has on the forms, the metal strips can be profitably used as shown in Figure 6.112.

Reshoring

Reshoring is a very simple operation, but one that also requires close monitoring of the growth of the concrete strength to ensure that the safety and serviceability of the concrete structure is not detrimentally affected.

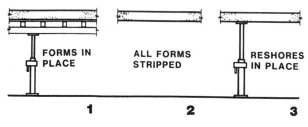

Figure 6.113 - THE RESHORING PROCEDURE

It involves the complete strip of the formwork followed by the immediate installation of reshores (props) at some specified maximum spacings and to a regular pattern. (Figure 6.113) The area being completely stripped may be only one bay at a time or can be the complete floor area. For obvious safety reasons, this matter must be under the control of the design engineer for the concrete structure.

This complete strip, at which the slab assumes all its own weight and that of any superimposed loads, often takes place at a very early age, frequently less than seven days. Therefore, it is imperative that the concrete mix is designed for early strength gain, and that this growth of strength be closely monitored and confirmed by tests before each act of reshoring is commenced.

Although the concrete has attained sufficient strength to be free spanning at a young age it will suffer from excessive creep deflection. Creep deflection occurs slowly over time and is greater with young concrete. This is the reason for the reshores. They prop the new slab down onto the area that supported the original formwork. Thus the creep deflection is greatly inhibited.

If any superimposed loads, formwork or materials, are placed on the new slab, the reshores transmit these loads down to the lower areas. The reshores remain in place until the superposed loads are reduced and the concrete is more mature. The time for this is also controlled by the design engineer for the concrete structure.

General Techniques for Removing Components.

Reference was made previously to the need to avoid physical shock and for the gradual transfer of load to the structural concrete. Further, it must be possible to remove the formwork components with the absolute minimum of damage. The better the condition of the components, the more often they can be reused, and the lower the formwork costs will be.

The provisions on propping to maintain the safety of the reinforced concrete structure have been covered with the description of the three sequences of stripping. This section covers the removal of the components: plywood and framing.

Firstly, the soffit formwork framing should be lowered 50 mm or more. Ideally, the formface plywood will drop with it. In practice if the metal strips shown in Figures 6.107 and 6.112 have been used they will readily drop down onto the framing, usually with a peeling action that starts at the metal strips.

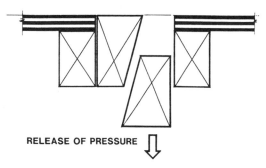

RELEASE OF PRESSURE

Figure 6.114 - RELIEVING THE FORMFACE - 1

A detail of another method for relieving the formface is shown in Figure 6.114. Here two tapered timbers provide a strip of the formface. After the bearer line has been lowered, one of the tapered timbers can be pulled down, usually by eye bolts in its soffit, and the pressure is relieved. After all the joists have been lowered onto the bearers, the resultant gap provides a start point to peel down the plywood.

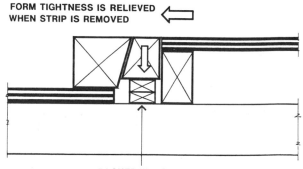

FORM TIGHTNESS IS RELIEVED WHEN STRIP IS REMOVED

PACKER WEDGES

Figure 6.115 - RELIEVING THE FORMFACE - 2

Where the soffit formface has a small step, this can be used for a tightness relieving detail as shown in Figure 6.115.

If methods to ease the tightening action of concrete shrinkage and grout penetration between plywood sheets have not been used, then the plywood will often have to be prised off the concrete soffit. The result is edge damage to the plywood and a severe reduction in its useful life.

With the plywood now resting on the joists, its removal can commence. At the simplest, it can be slid out from the edges of the form. If the slab is surrounded by beams, the joists near the middle of the form can be pushed to one side and the plywood sheets passed out from there.

Following this, the joists can be removed and then the bearers. If telescopic steel props have been used for supports, they will become unstable as the joists and bearers are being removed. Care will be required when removing them with the framing.

The essential point is that neither materials nor equipment should be allowed to drop the full form height during stripping. The impact on the ground or concrete slabs can break the corners of plywood sheets, crack or split framing timbers and damage the equipment.

The most potentially hazardous damage is that done to the timber framing members. Often this damage is not seen, and the defective and seriously weakened timbers are re-used. When loaded they can be the start point for a collapse.

ALUMINIUM BEAMS.

The proprietary aluminium beams used for soffit form framing are the same products that were described in Chapter 4 for use in wall forms. Figure 6.116 shows the cross section of two brands of aluminium beams.

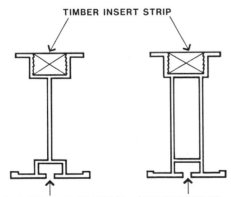

Figure 6.116 - TWO BRANDS OF ALUMINIUM BEAMS

When used as joists the timber nailing strips provide a ready fixing line, and the slot for 'tee' headed bolts on their lower flanges can be used for bolting them to steel beam bearers. Figure 6.117 shows two views of this.

The 'tee' headed bolts and the clips can be installed at any time after the aluminium beam has been placed in position. The head of the bolt is inserted into the groove in the lower flange and rotated through ninety degrees. The clip can then be positioned and the bolt tightened.

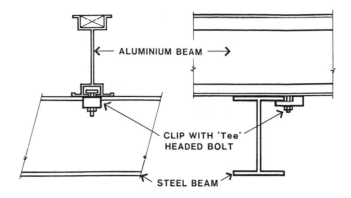

Figure 6.117 - ALUMINIUM BEAMS ON STEEL BEARERS

If an effective fixing is needed to a timber bearer, a saddle clip or bolted angle can be used. (Figure 6.118)

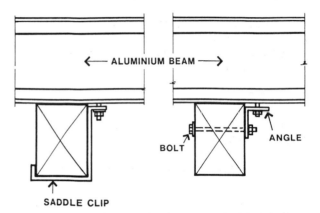

Figure 6.118 - FIXING ALUMINIUM BEAMS TO TIMBER

Where aluminium beams are used as bearers the joists can fixed by skew nailing to the nailing strip. The width of the aluminium bearers is usually less than 100 mm and the normal techniques of lapping the bearers over the supports can be used.

FLOOR CENTRES.

Floor centres are proprietary telescopic steel beams that are designed to provide adjustable-length framing members for soffit forms. Each manufacturer of these units provides data on the slab thickness-span-centre spacing relationship for each type of unit.

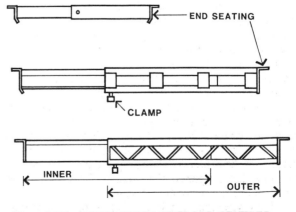

Figure 6.119 - THREE TYPES OF FLOOR CENTRES

Figure 6.119 shows three types of floor centres. The short span centres (up to 2.1 metres span) are often made of folded heavy gauge sheet metal while longer span centres (up to 6.2 metres span) are framed from steel rod and channel sections. Depending on the slab thickness and span, the recommended spacings vary from 200 mm to 1500 mm.

The length of the centre (the span) is adjusted by sliding the inner within the outer. This length is fixed by a clamping fitting on the lower flange that, for the longer centres, also locks the camber at its set value. Both ends have seating lugs for their support on beams, concrete walls or hangers.

As shown in Figure 6.120 the formface, plywood sheets or proprietary plywood-faced steel-edged form panels, sits directly on the floor centres. These proprietary panels are the same as those used in wall forms. (refer to Figure 4.153)

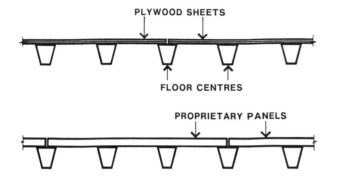

Figure 6.120 - SECTION OF FLOOR CENTRES

Steel floor centres do not have very much torsional resistance to the twisting effect of eccentric loading. Figure 6.121 illustrates an extreme example of this type of eccentric load effect.

If the progress of the pour comes from the left hand side, and a number of adjacent panels or plywood sheets all butt to the next sheet near the same edge of this centre, then the twisting action will accumulate along the centre and may fail it.

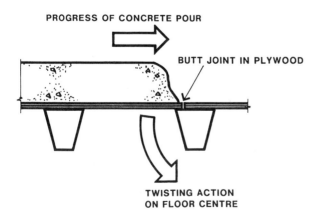

Figure 6.121 - TWISTING LOAD ON A FLOOR CENTRE

This situation can be avoided by ensuring that adjacent rows of panels or plywood sheets have their laps on different centres. Figure 6.122 is a part plan illustrating these staggered laps.

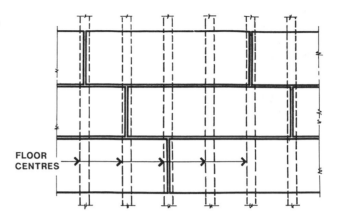

Figure 6.122 - PART PLAN SHOWING STAGGERED LAPS

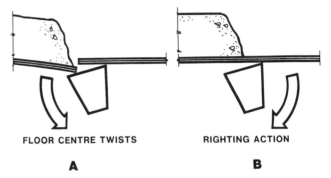

Figure 6.123 - THE TWISTING & RIGHTING ACTION

If one panel tends to eccentrically twist a floor centre it will rotate as shown in Figure 6.123 'A'. However, the adjacent panel, being continuous will effectively twist the centre the other way by bearing on the far edge as that edge lifts up. (Figure 6.123 'B') This righting action will tend to restore the floor centre's stability.

End Support of Centres

Where centres support the formwork for a slab spanning between reinforced concrete walls, the centres can be carried on the wall. Figure 6.124 shows a method that uses a typical type of beam hanger.

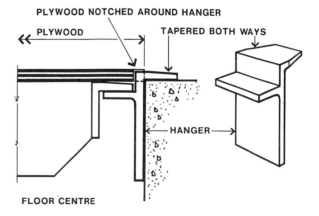

Figure 6.124 - FLOOR CENTRE SEATED ON BEAM HANGER

The top leg of the steel hanger sits on the top of the wall and the centre sits on the bracket to the hanger. Note that the plywood thickness must be great enough for the top of the plywood to be above the top of the hanger leg. This gives clearance for the tapered top leg of the hanger to be moved out when stripping the forms. Also, the edge of the plywood sheet is notched at the hangers to limit the grout loss at the junction of the plywood and the wall.

The end lugs of the centres are also tapered, and they can be seated directly on the top of the concrete wall as shown in detail 'A' of Figure 6.125

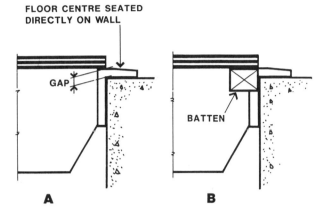

Figure 6.125 - CARRYING THE CENTRE ON THE WALL

This places the line of the plywood above the wall and to prevent grout loss this gap must be closed. Figure 6.125 'B' shows a timber batten fixed to the plywood.

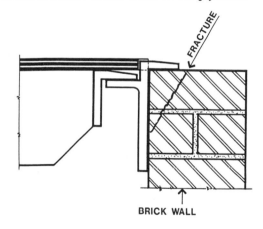

Figure 6.126 - FRACTURE FAILURE OF BRICK WALL

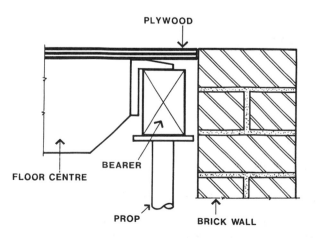

Figure 6.127 - BEARER SUPPORT BESIDE BRICK WALL

There has been a history of failures, like that shown in Figure 6.126, when either centres or their hangers have been seated on masonry (brick or concrete block) walls. Because of this it is not recommended and in some areas it has been prohibited by regulation.

When the walls are masonry, the formwork should be carried on bearers and supports. Figure 6.127 shows this arrangement.

Increasing the Load Capacity

An examination of the manufacturer's data on the slab thickness-span-centre spacing relationship will show although the floor centres can extend to quite long spans, they cannot carry thick slabs at those lengths. Provided certain precautions are observed, and the manufacturer's directions closely followed, the centres can be propped at mid-span and the permitted slab thickness thereby greatly increased. Figure 6.128 shows the side elevation and section of this arrangement.

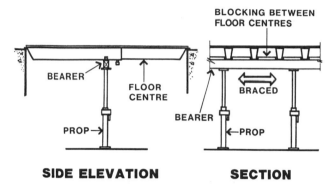

Figure 6.128 - MID-SPAN SUPPORT TO CENTRES

A suitable bearer is propped up against the underside of the centres. The actual position of the bearer on the centre must be located to the manufacturer's directions. Further, any lateral (sideways) movement in the bottom of the centres must be prevented. As shown, the bearer must be braced against movement in the direction of its length and the timber blocking, nailed to the top of the bearer, prevents the bottom of the centres buckling sideways.

Minor Setdowns in Soffit Forms

Because of the ease with which the length of the floor centres can be adjusted, they are very suitable for forming minor setdowns in the slab soffit. Figure 6.129 shows a simple detail of a section through the formwork step.

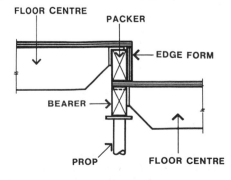

Figure 6.129 - MINOR SETDOWN OVER BEARER

A continuous packer, of the appropriate height is seated on the lower form plywood and directly above the bearer to the lower area. This method is the recommended one.

However, if the packers are located out on the span of the lower area, structural failure of the lower centres can occur due to the high shear forces from the point loads. As Figure 6.130 indicates, although the bending effect of this arrangement may be within the capacity of the centres, the shear forces are often excessive.

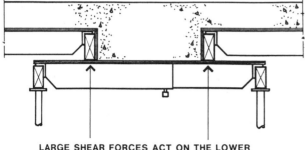

LARGE SHEAR FORCES ACT ON THE LOWER
FLOOR CENTRES AT THESE POSITIONS

Figure 6.130 - POSSIBLE SHEAR FAILURE OF CENTRE

There has been a history of failures of centres from this method of framing and it is not recommended. Some simple additional work can make the arrangement safe.

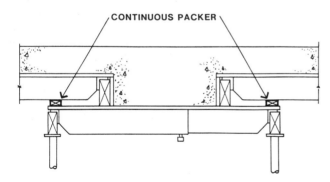

Figure 6.131 - PACKERS TO AVOID SHEAR FAILURE

Figure 6.131 shows continuous packing between the bottom of the upper centre and the lower form, directly over the lower bearer. This effectively eliminates the excessive shear forces, from the upper centres, that act on the lower centres.

Eccentric Frame Loading from Centres

It was noted earlier that a range of bracing lengths was available for frames, giving spacings up to 3.0 m. It has become common practice to use floor centres to span this 3.0 m with the bearers spanning the 1.2 m across the frame. This is shown in Figure 6.132.

It will be noted that this places any eccentric loading from the bearer in the weakest direction of the frame. Figure 6.61 illustrated this earlier in this chapter. Such a loading arrangement can reduce the load capacity of the frames by more than 50% below their maximum. If this method of forming is adopted the loads on the frames should be carefully checked against the permitted values given in the manufacturer's literature.

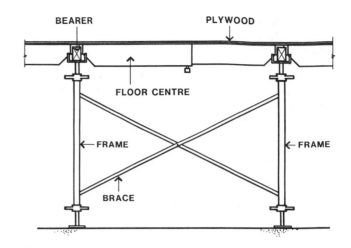

Figure 6.132 - LOADING IN THE WEAKEST DIRECTION

Framing the centres the other way, with respect to the frames, can also cause eccentricity problems. Figure 6.133 shows this.

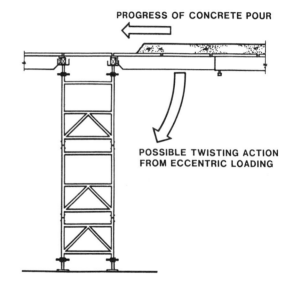

Figure 6.133 - ECCENTRIC LOADING ON TALL FRAMES

If the loading is mostly from one side, and the bearer is eccentric on the top of the frame then there will be severe twisting loads on the multi-storey frame towers.

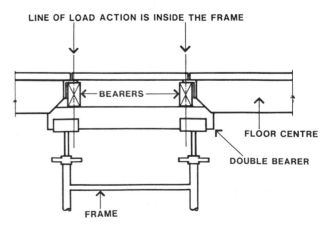

Figure 6.134 - ADDING ANOTHER LEVEL OF FRAMING

By adding another level of timber framing, the loads can be made to act within the frame and the twisting action will be reduced.

Figure 6.134 shows short bearers across the frame in the 'U' heads. Double bearers are used to ensure that they are always as concentric as possible. Above these, the bearers carrying the floor centres, span at rightangles to the frames, and are located a short distance inside the frame legs.

By ensuring that the load action is located inside the frame supports, the effect of the eccentric loading is minimised.

Stripping Floor Centres

Figure 6.135 gives a diagramatic representation of a floor centre and the soffit forms. The numbers represent the order of the sequence of stripping.

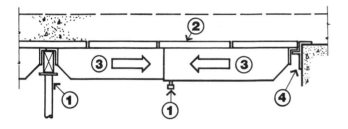

Figure 6.135 - FLOOR CENTRE STRIPPING SEQUENCE

The sequence of stripping a floor centre is given in four stages:

1. The camber lockbolt is released. The ends are lowered if possible.
2. The form panels or plywood sheets are removed.
3. The floor centre is telescoped inwards to be clear of its supports.
4. Wall hangers, if any, are removed.

If any backpropping or reshoring is specified this can be done in conjunction with this procedure. Backpropping and stripping bands can also be incorporated.

QUICK STRIP SYSTEMS.

Essentially, quick strip systems are proprietary hardware that conforms to the requirements for undisturbed supports that were described earlier in this chapter. Two typical types will be described.

The first comprises a special set of steel framing members that are attached to the top of modular framing systems. The second uses aluminium structural members and has the unique feature of being able to be substantially erected from the footing level.

At the top of the standards (props) of the braced and tied modular support system, in the first example, a special soffit support and lowering mechanism is fitted. Two examples of this fitting are shown in Figure 6.136

These fittings to the top of the standards have the following features in common:

1. The cap plate is part of the formface and remains

in contact with the concrete.
2. There are side brackets to carry the soffit form framing members.
3. The side brackets are attached to a cam operated lowering device which has two actions: the first is for a slow lowering for approximately 25 mm (gradual transfer of load during stripping), the second gives a more rapid drop for approximately a further 50 mm.

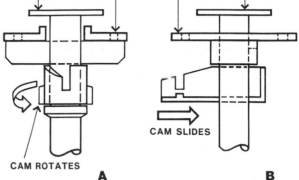

Figure 6.136 - EXAMPLES OF TOP OF PROP FITTINGS

Detail 'A' of Figure 6.136 shows a rotary cam for the first slow movement. At the end of this action, the side bracket drops rapidly into a slot for a further 50 mm. Detail 'B' has a linear cam (inclined plane) for the first slow lowering movement. This is also followed by a drop.

Framing members are seated on these side brackets to give continuous lines of support. To avoid dislodgement, the ends of the framing members are usually pegged into the bracket. Figure 6.137 shows an elevation of one type of framing member, a lattice beam. In this example, the tops of the framing members are also the formface.

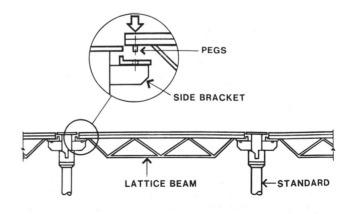

Figure 6.137 - ELEVATION OF FRAMING MEMBERS

Deck frames span between the line of framing members with plywood sheets on top of the deck members. In most cases the plywood is not fixed. In the example shown in Figure 6.138, the system is arranged so that standard plywood sheets will fit neatly between the top edges of the framing members. This avoids any wasteful cutting of plywood sheets.

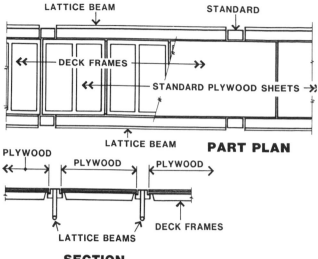

Figure 6.138 - DECK FRAMES AND PLYWOOD FORMFACE

After the Removal of All Formwork Framing and Plywood

As noted above, the second example uses aluminium framing members. Similar to the first case a special support mechanism is fitted to the top of the supports. However, in contrast to the first system, the plywood sheeting passes over the top of the cap plate of the special fitting. As shown in Figure 6.139, primary aluminium beams hook over lips on the sides of a mid-height support bracket to the special fitting.

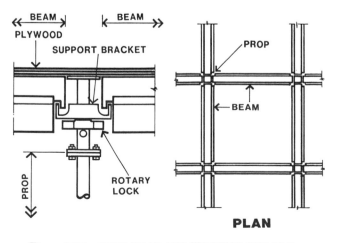

Figure 6.139 - PROP HEAD AND FRAMING DETAILS

These primary aluminium beams frame between all props in both directions and give them stability. The lowering mechanism is a rotary lock just below the plate supporting the beams.

Secondary aluminium beams frame between the primary beams. They sit on a continuous lip which is provided on both sides of the primary beams. Both the primary and the secondary aluminium beams have a continuous timber insert in the top for the nailing of the plywood formface if required. The secondary beams are not shown in the plan view of Figure 6.139.

The tops of the special fitting cap plates and the tops of both primary and secondary aluminium beams are all at the same level. Plywood sheets are placed on them. Lacking the retaining edges that the first system had, it is necessary to nail the sheets at the periphery of the formed area. With the outer sheets nailed there will be little chance for the plywood to move.

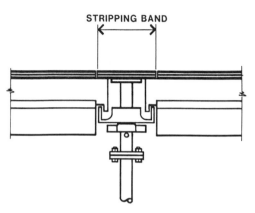

Figure 6.140 - STRIPPING BANDS OVER THE PROPS

For the system to act as an 'undisturbed support' formwork arrangement, narrow plywood stripping bands must be placed over the lines of props. (Figure 6.140) As described previously, when the soffit formwork is lowered and removed, these strips and the props under them remain in place.

The erection procedure for this type of system has the unique advantage that the props, and the first row of primary beams, can be erected from below. Figure 6.141 shows the sequence.

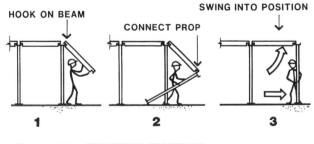

Figure 6.141 - ERECTION SEQUENCE

To add another primary beam, the formworker works from below the formwork. It is hooked onto the supporting lip of the end standing prop. The next prop can then be hooked onto the outer end of the beam and, still working at the lower level, the beam and the prop can be swung up into position.

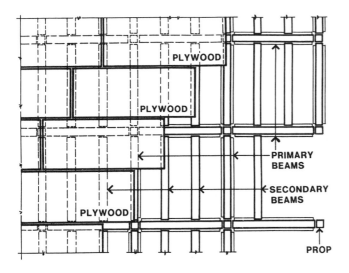

Figure 6.142 - PLAN OF SYSTEM DURING ERECTION

Working from the top, the secondary beams can be placed between the primary beams and, if required pushed out into position. (Figure 6.142) As the installation of the secondary beams is completed, the plywood is placed to provided a safe working area up at the edge of the work.

The stripping of the soffit form framing members is done in a manner similar to the other forms.

CHAPTER 7: BEAM FORMS

In a typical reinforced concrete building frame, the slabs, which are formed with the soffit forms described in the previous chapter, are the primary load carrying element. They transfer their load to secondary elements, such as walls or beams. This chapter is concerned with the second example, the forming of beams.

Beams can be grouped in two general types. When located in the interior of the structure they are usually TEE beams and 'L' beams when on the perimeter. Figure 7.01 shows typical examples of these.

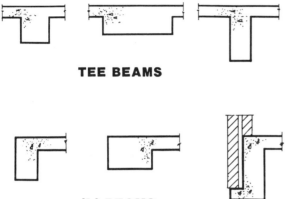

TEE BEAMS

'L' BEAMS

Figure 7.01 - EXAMPLES OF BEAMS

In both cases, widths and depths can vary greatly. The shallower beams can often be formed as minor set-downs in the soffit form. This case was covered in Chapter 6. Deep beams are effectively short walls, and some of the techniques given in Chapter 4 are used for them.

All the loadings described for soffit forms act on beam forms. Also, the basic structural principles of **STABILITY, STRENGTH** and **SERVICEABILITY** apply throughout the three stages of construction. The vertical loads of live load, concrete load and formwork self-weight are diagramatically shown acting on the beam formwork in Figure 7.02. Because of the greater concrete depth, the pressure on the bottom of the beam form is greater than that on the slab.

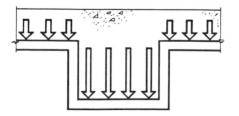

Figure 7.02 - BEAM FORM VERTICAL LOADING

In most cases, some portion of the vertical loads that act on the soffit forms are carried on the sides of the beam form.

The horizontal loads acting on the form come from two sources: concrete pressure on the beam sides and the horizontal loads acting on the total formwork system. Figure 7.03 illustrates the horizontal concrete pressures. These forces are resisted by either bracing the beam sides, or tying them together in a manner similar to wall forms.

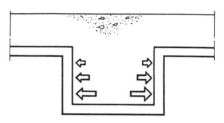

Figure 7.03 - HORIZONTAL CONCRETE PRESSURES

The horizontal loads that act on the total formwork system comprise wind loads, construction loads and impact. As Figure 7.04 shows, the beam forms are often required to transmit these externally applied loads across the beam form, and distribute them to the remainder of the form and the bracing systems.

FORCE TO BE TRANSMITTED ACROSS BEAM FORM

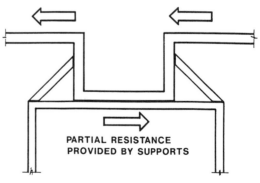

PARTIAL RESISTANCE
PROVIDED BY SUPPORTS

Figure 7.04 - TRANSMITTING HORIZONTAL FORCES

The most common method of achieving stability under these loads is bracing to the beam sides with adequate fixings between the soffit forms and the beam sides.

Another stability problem, which can also have catastropic consequences, can occur with the selection of the beam form support method. All formwork is progressively loaded as the concreting proceeds. Wall and column forms are filled starting at the base and the fluid concrete pressures increases as the pour proceeds.

With horizontal formwork, soffits and beams, the concrete pour starts at one point on the form, usually at one side, and proceeds across the form from there. Parts of the formwork, which will be symetrically loaded when completely concreted, are often eccentrically loaded as the pour progresses. Beam forms carried on a single support are one such case.

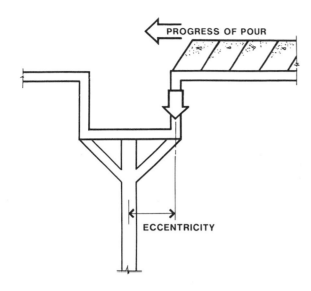

Figure 7.05 - A DANGEROUS SINGLE SUPPORT

Figure 7.05 shows this support method. If the progress of the concrete pour is from one side, the form will be subject to a large eccentric load. This case was also covered in Chapter 1. It method is very dangerous and should not be used.

Another potentially dangerous case of asymmetrical loading is shown in Figure 7.06. A form for a very wide beam has been constructed on a support frame which is narrower than the beam. Some part of the soffit is carried on the edge of the wide beam form.

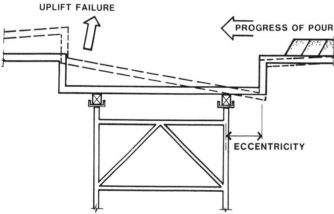

Figure 7.06 - AN UNSTABLE BEAM FORM

If the progress of the concrete pour is from one side, then the light weight of the empty formwork will usually be insufficient to counter the eccentric action of the concrete covered soffit form. Failure will result. For safety, the points of application of the loads must always lie inside the base of support.

The progress of a pour on such a formwork framing arrangement is shown in Figure 7.07. Although the loads, and the reactions to them, vary as the pour progresses, there is never any eccentric action that will de-stabilise the formwork. Because of the framing arrangement, the loads on the beam sides will always lie inside the supports. This is the general framing arrangement which is recommended.

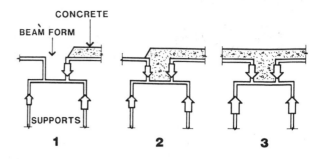

Figure 7.07 - LOADS ALWAYS INSIDE THE SUPPORTS

The load path within this type of formwork framing also merits examination. Figure 7.08 shows the slab and beam loads, and the path of these loads down to the supports of the formwork system. The beam sides transmit loads from the soffit form down to the beam form framing.

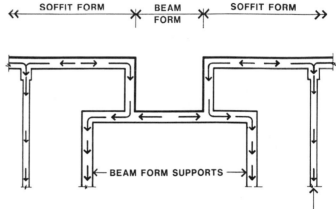

Figure 7.08 - LOAD PATH WITHIN THE BEAM FORMWORK

The bigger the span of the soffit form that is carried on the beam form sides, the greater is the bending effect on the framing carrying the bottom of the beam formwork. This is not structurally incorrect, but these loads can result in very large and uneconomic framing members. The bending effect is reduced if the span of the soffit forming onto the beam side is kept short.(Figure 7.09)

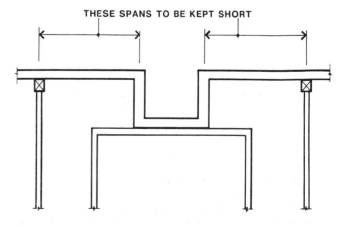

Figure 7.09 - MINIMISE THE LOADS ON THE BEAM SIDES

Another method that reduces this bending load is illustrated in Figure 7.10. The load path is shown on one half of the symmetrical formwork arrangement. Here struts, from the support legs up to an additional bearer, take the soffit form loading directly to the support leg. This reduces the bending load on the beam form framing members.

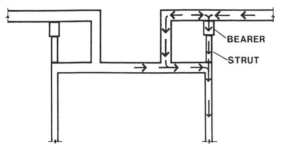

Figure 7.10 - REDIRECTING THE LOAD PATH

Construction details for these techniques for optimising the structural action are given later in this chapter.

GENERAL PRINCIPLES OF STRIPPING

Authorative text books and Codes of Practice on reinforced concrete work invariably call for slabs to be stripped before beams. The reason for this can be best explained by showing the problems that occur if the beams are stripped before the slabs.

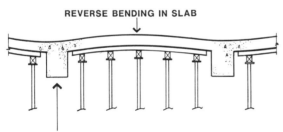

Figure 7.11 - BEAMS STRIPPED BEFORE SLABS

All structural members deflect when loaded, and this includes the state when only carrying their own weight. Sometimes this deflection is very small, but it occurs nevetheless. Figure 7.11 diagramatically illustrates a cross section through a beam and slab structure where the beams have been stripped first. The beams have deflected under their own weight but the slab, still being supported by its formwork, has not. The beams are partly carried by the edges of the slab forms and a reverse bending action has occured. This bending action is not compatible with the reinforcement pattern in the concrete. Unacceptable cracking can result.

When the slabs are stripped before the beams, the bending action on the concrete follows that designed for in the reinforcement pattern. The slab deflects to its expected shape, with its load being passed onto the beams, then the beam is stripped and deflects under load.

For the RESHORING procedure (described in Chapter 6) the slab is completely stripped first and then the beams are completely stripped. Following this the beams are reshored and then the slabs.

BACKPROPPING (also described in Chapter 6) starts with the backpropping of the beams and then the slab. At the second stage, the backprops to the slab are stripped first, to be followed by those to the beams.

The principle of designing the formwork so that parts of the formwork can be stripped out early has been mentioned in other chapters. The UNDISTURBED SUPPORT systems described in Chapter 6 can do this for soffits of both slabs and beams. Further to this, it is possible to strip the beam side formwork without disturbing the soffit. Some details of formwork that enable this to be done are given later in this chapter.

CONSTRUCTION DETAILS

Formwork to the Base of Beams.

Two aspects of the base framing are significant: the width of the formed face and the direction of the framing members. On the first matter, many textbooks show details of beam formwork with the formed base only as wide as the concrete beam itself. Figure 7.12 is an example. The beam sides and soffit forms are only shown in outline.

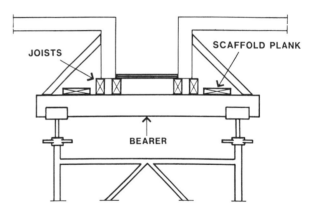

Figure 7.12 - NARROW BASED BEAM FORM

While this may seem to be a neat and obvious solution to the task, it has some deficiencies. Firstly, there are difficulties in attaching the braces to the beam sides, they have to be on the line of the bearers which are often quite widely spaced. If more closely spaced braces are needed then a bearer sized member is needed along the line of the base of the braces.

Secondly, the safe access for formworkers is poor and usually only consists of a single line of scaffold planks on each side of the beam form. Poor access leads to low productivity and increased costs.

Finally, the plywood for the base of the beam form has to be cut to width. This markedly reduces its useful life and possible number of re-uses. Again, an increase in costs.

Figure 7.13 shows the cross section of a wide based beam form that avoids these access and plywood cutting problems.

The principal feature is the line of full-width uncut standard plywood sheets which are the base of the beam form, and provide a work area. As uncut plywood sheets, their useful life is greatly extended.

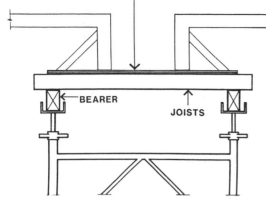

Figure 7.13 - WIDE BASED BEAM FORMWORK

The full-width plywood sheets also provide continuous tie lines across the form for the resolved forces from the braces and the concrete pressure. (Figure 7.14)

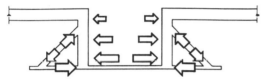

Figure 7.14 - CONCRETE PRESSURES AND REACTIONS

To cater for these forces, the braces can be installed at relatively close spacings if needed. If the framing consists of transverse joists and bearers parallel to the beam, braces can be fixed above any of the joists. This type of base framing is shown in Figure 7.15.

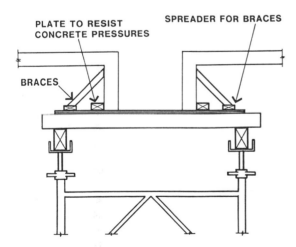

Figure 7.15 - BRACES TO BEAM SIDES

If the chosen spacing for the braces does not suit the joist spacing, they should not be fixed to the plywood at points between the joists. A spreader should be installed to take the thrust from the braces.

Figure 7.15 also shows a plate fixed to the base to resist concrete pressures on the beam sides. While this is not essential, the bottom plate to the beamsides can be fixed directly to the base, it does makes the accurate set-out of the beams easier.

The framing system shown in Figure 7.15 has the good feature of framing the bearers in the best direction to suit the strength of the support frame.

However, this framing method has one deficiency. As the cut-away perspective view of Figure 7.16 shows, the span of the plywood is parallel to the beam form sides, and the normal deflection of the plywood under concrete pressure will lead to grout loss.

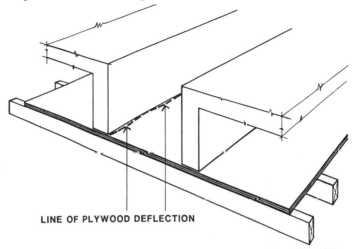

Figure 7.16 - DEFLECTION OF PLYWOOD OF BEAM BASE

If it is important that this grout loss be prevented, then the formwork framing shown in Figure 7.17 should be used.

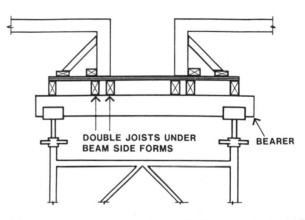

Figure 7.17 - FRAMING TO CONTROL PLYWOOD DEFLECTION

Here the plywood spans transversely with two longitudinal joists under each beam side. These prevent grout loss through plywood deflection under the beam sides. The full width plywood still provides the tie member for the bracing action, and the braces may be fixed to the plywood anywhere on the line of the outer joists.

One potential problem comes from the direction of the bearers. As explained in Chapter 6, this can lead to the worst case for eccentric loading of the support frames. Double bearers can be used as an aid to achieving concentric action load action on the frame, but this is still the frame's weakest loading direction. If the matter of the frame's load capacity is considered to be critical, an extra level of bearers can be used to alter the framing direction at the top of the frame. This is illustrated in the section of Figure 7.18.

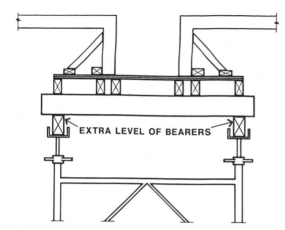

Figure 7.18 - ALTERING THE FRAMING DIRECTION

For a wide beam the support base must also be wide. Figure 7.19 shows the cross section of one of these beam forms. The support frames are selected to be wider than the beam so that the loads on the beam sides will always lie within the outer frame supports. The framing is two bays of modular frames.

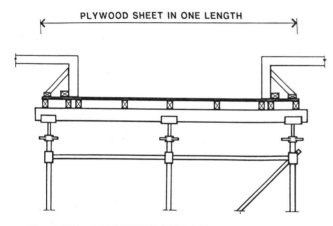

Figure 7.19 - FORMWORK FOR A WIDE BEAM

In this case, the plywood sheets are placed with their length across the beam. This provides the necessary continuous connection across the form to cater for the forces from the braces and the concrete pressure on the beam form sides.

Beam Form Sides.

The desireability of limiting the load that acts on the beam sides from the soffit forms was mentioned earlier. One of the extreme cases of such loading is the support of telescopic floor centres on the beam sides. Figure 7.20 shows the cross section of an example of telescopic centres spanning between beam forms. To carry the vertical loads, and also be able to resist the concrete pressures, the beam sides are framed with top and bottom plates and studs. The spacing of these studs does not usually exceed 600 mm.

Cases like this occur when there is a need to provide clear access under the formwork for other site workers and movement of materials. As the Figure shows, much of the load from the centres, that would go onto the beam sides, can be diverted directly down to the supports by providing

extra joists to the edge of the form base and packing between the centres and the form base.

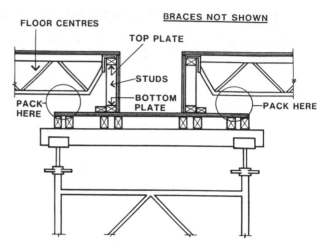

Figure 7.20 - FLOOR CENTRES SEATED ON BEAM SIDES

This packing, shown in Figure 7.21, must be done carefully. If it is too tight, then, when the framing under the beam form deflects due to the weight of the concrete in the beam form, the ends of the floor centres will lift off the top of the beam sides. Without the sideways restraint at its top that this seating provides, the centres may roll over. Thin spacer strips fixed to the top of the beam sides, between the centres, will help prevent this.

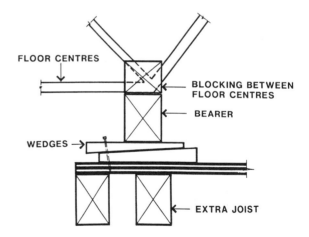

Figure 7.21 - PACKING UNDER FLOOR CENTRES

The point of application of the packing to the floor centre is also very important. As Figure 7.21 indicates, the packing should be placed under a panel point of the framing of the floor centre. In each case, the formworker should follow the manufacturer's directions on the recommended installation method and position of this packing.

The bottoms of the floor centre at the packing also require lateral restraint. As shown, blocking between the centres, which is fixed to the packing members, provides this.

Figure 7.22 shows the cross section of a beam form with the joists seated on and fixed to the beam sides. As indicated, strutting can be placed, between the joists and the beam form base, to relieve some of the load that would otherwise be totally carried on the beam sides.

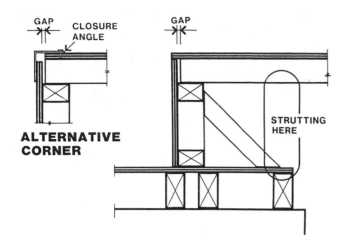

Figure 7.22 - JOISTS SEATING ON THE BEAM SIDES

Alternatives are given for the detail at the top of the beam sides. In the plywood to plywood corner junction, a gap is left at the end of the joist to ease its stripping. However, some jambing of the plywood can still occur.

The alternative detail gives a solution to this. A sheet metal angle, usually 0.8 mm thick, closes off the small gap between the ends of the two plywood faces. To prevent movement of the angle during concrete placement, it is shown fixed with short clouts to the soffit form plywood. This angle also provides a tolerance gap for the control of the width of the soffit formwork. (Refer to Chapter 6 Figures 6.106, 6.107 and 6.108)

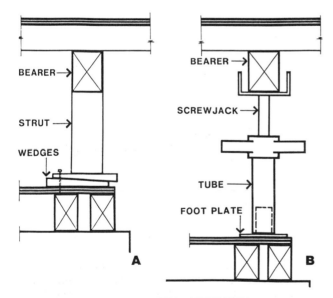

Figure 7.23 - STRUTTING THE SOFFIT FORMS

Two methods of strutting are shown in Figure 7.23. The left-hand example consists of a bearer under the soffit form joists with timber struts on wedges. The struts are located immediately above the legs of the support frames. The right-hand detail can be used if there is sufficient height available to fit in the combined height of the bearer, adjustable 'U' head, tube and spigotted footplate.

In both cases the struts are located over two or more joists. These assist in minimising any eccentric load effects that might act on the support frames from misalignments of the strutting.

Another common technique for limiting the load of the soffit acting on the beam sides is to provide separate support systems for the soffit forms and the beam forms. Figure 7.24 shows sectional views of the two associated beam side to soffit form junctions: one at right-angles to the soffit form bearers and the other parallel to them.

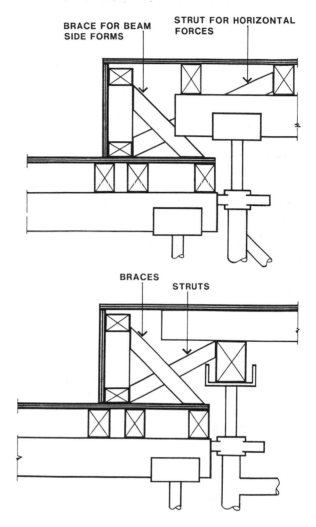

Figure 7.24 - SEPARATE SUPPORTS FOR BEAM AND SOFFIT FORM SYSTEMS

In both cases, the only load on the beam sides is from the narrow strip of plywood that spans between the two systems. To ensure that no instability can occur during the progressive loading of the soffit forms, the cantilevers of ends of the bearers and joists must be kept to a safe minimum.

Consideration must also be given to the ability of the formwork to transmit horizontal forces across its construction. (refer to Figure 7.04). The plywood to plywood junction shown in Figure 7.24 is not adequate for this. The detail shows additional struts from the beam form up to the soffit form framing to transmit the horizontal loads. These struts are not a substitute for braces to the beam sides. They are also needed; both are shown on the detail.

Attention is drawn to the details the support of the soffit form plywood on the beam sides. It is constructed this way to minimize any grout penetration under the plywood.

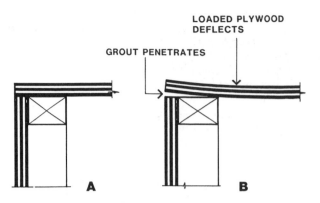

Figure 7.25 - GROUT ENTRY UNDER THE PLYWOOD

If the plywood edge were to be supported as shown in Figure 7.25 'A' then the deflected shape shown in Figure 7.25 'B' will occur as the concrete is placed. Grout will penetrate under the plywood, leading to moisture loss, boney concrete and hydration staining. Further, stripping is more difficult.

The details of Figure 7.24 can be developed further to enable easier stripping of the formwork. Figure 7.26 shows a simple amendment to Figure 7.24.

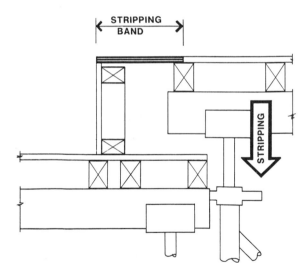

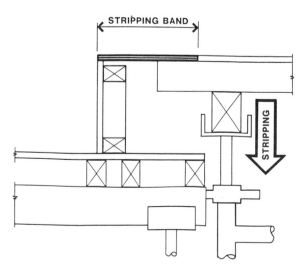

Figure 7.26 - STRIPPING BANDS AT BEAM SIDES

Similar to the purpose of the stripping bands described in Chapter 6, narrow strips of the plywood at the edges, provides stripping bands that enable the easier release of the soffit forms before the beam forms.

By adopting the metal closure angle shown previously, the details can be further simplified and improved as shown in Figure 7.27.

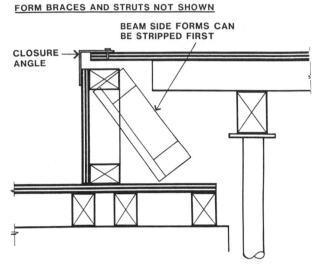

Figure 7.27 - TOLERANCE GAP WITH CLOSURE ANGLE

The use of the closure angle for a tolerance gap also permits the stripping of the beam sides without disturbing either soffit forms or beam base forms. After the removal of the struts, braces and bottom fixings to the beam side, they can be removed. For multi-storey buildings or other repetitive work this can give cost advantages.

Deep Beams

The special feature of formwork for deep beams is the need to cater for the greater horizontal concrete pressures that will occur. Tie rod systems from wall formwork hardware are used. To aid the control of the beam width, only tie rod systems that both hold the formfaces apart and together should be used.

Where a single line of tie rods is to be used, special attention must be paid to the fixings of the beam sides to the beam base. To illustrate the need for this, Figure 7.28 shows the change in forces and reactions at the base, as the concrete pour proceeds up the beam.

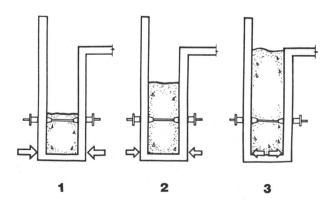

Figure 7.28 - THE CHANGING FORCES AT THE BASE

At the early part of the beam filling, the concrete pressure forces the bottom of the beam sides apart and the fixings must hold them in. With deeper concrete, this force diminishes and then reverses when the beam is nearly full. From that point on, the bottom fixings are required to hold the bottoms of the beam sides apart. With two rows of ties this reversal of reaction is rarely a problem.

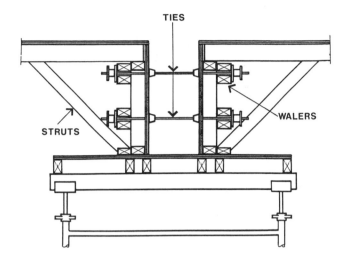

Figure 7.29 - DEEP BEAMS WITH TWO ROWS OF TIES

Figure 7.29 shows a typical cross section of a deep beam with two rows of ties. Note that braces are not needed to resist the concrete pressures, the ties do that, but struts are needed to transmit any other horizontal forces across the beam form. These struts also maintain the plumb of the beam sides.

Internal Beams of Differing Depth.

The intersection of the formwork for the junction of two concrete beams of differing depth can complicate the formwork. Indeed, one of the established recommendations for the minimisation of formwork costs is for all concrete beams to have the same soffit-level.

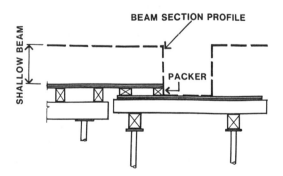

Figure 7.30 - JUNCTION OF DIFFERENT FORM LEVELS

As shown in Figure 7.30, the plywood base to the forms of the shallower beam laps over that of the deeper beam. Note that this plywood extends up to the face of the deeper concrete beam, and the packer on the end is fixed on the same line. The end of this upper plywood sheet is cut to fit around the ends of the beam form sides of the deeper concrete beam. Figure 7.31 gives a perspective view.

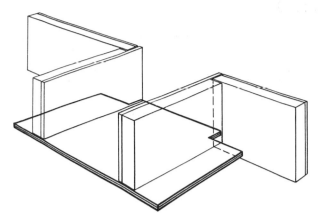

Figure 7.31 - END CUTTING TO BEAM SOFFIT FORM

The plywood face of the shallower beam form side can then lap past the end of the deeper form. Figure 7.32 shows a perspective view of this arrangement with a plan-section of the corner junction.

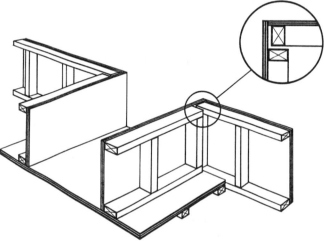

Figure 7.32 - JUNCTION OF BEAM SIDE FORMS

To prevent the beam sides being locked in at their ends by a tight interface with the concrete, a sheetmetal closure angle can be used. Figure 7.33 shows a plan-section.

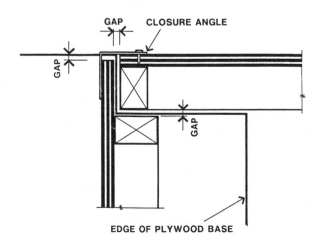

Figure 7.33 - CLOSURE ANGLE AT BEAM SIDE JUNCTIONS

Edge Beams.

A typical detail of an edge beam ('L' beam) is shown in Figure 7.34. The important factor to cater for here is the need for stronger braces on the outer beam side form to resist the pressure from the deeper concrete. Horizontal ties, from the beam form supports to the soffit forms, are usually needed to hold the two formwork systems together. These are not shown in this detail.

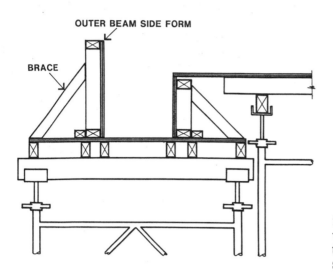

Figure 7.34 - TYPICAL EDGE BEAM FORMWORK

The outer form is often further complicated by the need for a concrete ledge at the base to carry brickwork or other external cladding. A cross section through the outer beam face formwork for such a case is given in Figure 7.35.

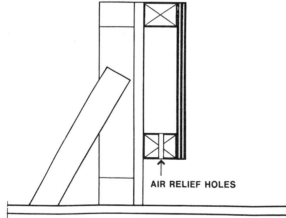

Figure 7.35 - OUTER FACE PACKER FORM

Two framed formfaces are involved, the outer one being full height. The setback for the beam face is formed with a plywood faced frame fixed to the outer formface.

Similar to haunches on walls, as described in Chapter 4, there is a need to permit air trapped at the top of the ledge to escape. Air relief holes should be drilled here at regular intervals. Grout will enter and clog these holes, and they need to be cleaned out at each re-use of the formwork.

Deep edge beams follow the construction of deep 'TEE' beams described above. Figure 7.36 shows the cross section of a deep edge beam with two rows of ties. In this

example, the outside walers are vertical and the formface has horizontal joists.

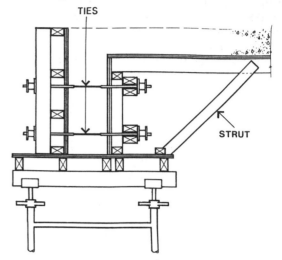

Figure 7.36 - SECTION OF DEEP EDGE BEAM FORMWORK

Alternatively, the outside forms could have been framed similar to the inner forms. These have horizontal walers and a stud and plate framed formface. To cater for the horizontal forces and maintain the plumb of the forms, struts are needed from the form base up to the framing of the soffit forms.

Edge Beams in Restricted Space.

Edge beams often have to be formed alongside the wall of a neighbouring property in a very small gap. The simple method of installing removable packing to form the gap, and the associated bracing problem, was indicated in Chapter 1 (Figure 1.04). In this method the concrete pressure acts on the wall of the next door building. In some cases it may not be capable of resisting it; in others it may not be permitted. To solve this problem of how to form the beam sides in a very narrow space, many methods have been devised on building sites. This section records some of them.

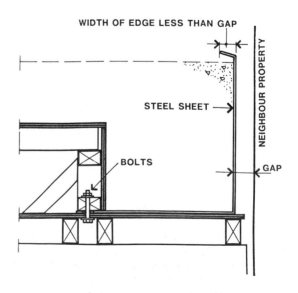

Figure 7.37 - FOLDED STEEL SHEET EDGE BEAM FORM

The section given in Figure 7.37 shows the cross section of lengths of folded mild steel sheet seated on the bottom of the plywood beam form and cantilevering up from there. The thickness of the steel sheet is calculated to be adequate to resist the concrete pressures and have a deflection within the concrete surface tolerances.

The inner edge of the folded sheet is anchored down by inserting it under the inner beam-side form. The fixings through the lower plate on the beam timber side (usually bolts) must be adequate to resist the uplift and the sliding action caused by the concrete pressure acting on the outer sheet steel beam face.

The sections of folded sheet are butted end-to-end, and if these joints were left unprotected, grout penetration would occur. This would make stripping very difficult. To prevent this and give ease of stripping, it is recomended that the units be placed with a small gap (1 mm) between them, and grout penetration prevented by installing a heavy duty adhesive tape across the butt joint.

The top edge is folded over to stiffen it, and thus maintain a straight line. So that the sheet metal form can be removed downwards and out under the beam, the top edge must be narrower than the gap between the beam face and the neighbouring property. If the gap is very narrow then a removeable stiffening lip can be provided. This is shown in Figure 7.38.

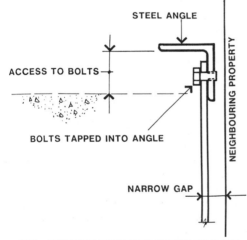

Figure 7.38 - REMOVEABLE TOP STIFFENING LIP

The angle is drilled and tapped (usually at 400 mm centres) so that the top edge of the form face can be bolted to it. The top edge of beam side extends above the concrete level to provide access for removal of the screws.

When these metal forms are stripped they have to be man-handled out from the bottom of the beam. This sets a limitation on the weight of the units. For 400 mm deep forms the steel sheet needs to be at least 4 mm thick; for 600 mm forms 6 mm plate is needed. Obviously this limits the depth of beam that can be formed. However, if the vertical cantilever of the sheet steel beam face can be tied at or near the top, then thinner and lighter steel plate can be used. Figure 7.39 shows the cross section of an example of this.

This method involves a row of ties as close to the underside of the slab as possible. This reduces the cantilever height to the practical minimum. Threaded rods, sheathed with plastic tubes, are used for the ties. They are anchored to the sheet metal beam side with nuts welded onto the outer face. Grout often penetrates into the tube

and makes the removal of the threaded rod very difficult. A heavy layer of grease on the rod will usually aid its removal.

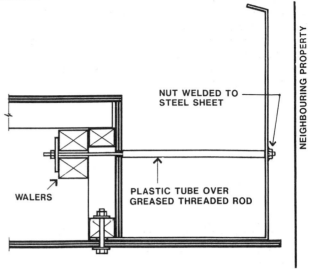

Figure 7.39 - TIES TO REDUCE THE CANTILEVER HEIGHT

Alternatively, the sheet metal beam face can be tied at the top using some components from wall tie systems. (Figure 7.40).

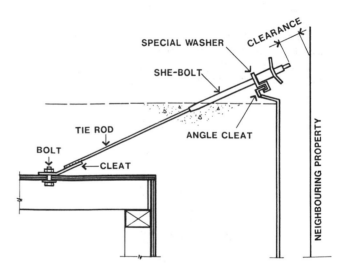

Figure 7.40 - TYING THE TOP OF THE EDGE FORM

This detail shows the top of the sheet steel beam face tied to the soffit formface at regular intervals. Angle cleats are welded to the stiffened top edge of the beam face at positions to suit the ties. A special washer on a She-bolt hooks over this cleat. The She-bolt connects to the tie rod which has been welded to an end plate or is fitted with an eye-bolt end. The bolting of this end to the soffit formface completes the tie assembly.

For ease of removal of the She-bolt, its point of connection to the tie rod should be very close to the surface of the concrete. If it were deeply embedded in the concrete there might not be enough space between the outer end of the She-bolt and the neighbour's wall, for its removal. Access from below is also needed for the removal of the bolt at the end of the tie rod, so that the soffit can be stripped.

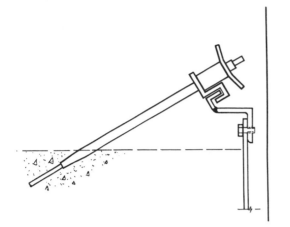

Figure 7.41 - ALTERNATIVE FOR REMOVABLE TOP EDGE

As Figure 7.41 shows the method using ties to the top can also be adapted to cases where there is only a narrow gap between buildings. The angle cleats are welded to the removable edge stiffening angle.

Edge Beam Stability - Multi-storey Work

Throughout this book the paramount formwork construction requirement of **safety and stability at all times** has been frequently emphasised. Another important case where this is a consideration is the formwork for the edge beams of floors to buildings of more than one storey. The stability problem occurs because the area to carry the formwork is no bigger than the area being formed. As a result, many of the loads lie outside the base of support. Figure 7.42 shows an edge beam form after concreting.

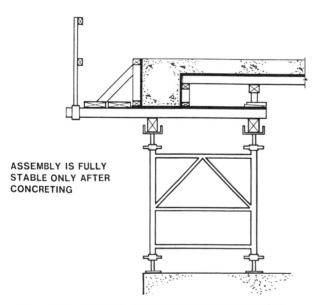

ASSEMBLY IS FULLY STABLE ONLY AFTER CONCRETING

Figure 7.42 - SECTION OF EDGE BEAM FORMWORK

With the heavy weight of the concrete, in addition to the light formwork weight, acting within and across the base of the support, the completed work in this case is stable. However, that does not mean that the formwork is stable before concrete placement. Figure 7.43 shows an

early stage in the construction of edge beam forms like those shown in Figure 7.42.

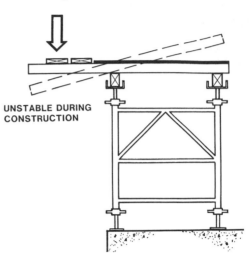

UNSTABLE DURING CONSTRUCTION

Figure 7.43 - UNSAFE START OF FORM CONSTRUCTION

At this time it is unsafe. Any load applied at the outer end of the cantilevering beam would cause the beam to tip. The beam can be made safe by adequately propping its outer end. The prop, cantilevering beam, bearers and frames must be fixed together. However, if this propping is installed as shown in Figure 7.44, a sufficiently large force at the outer edge will tend to cause the whole assembly to overturn.

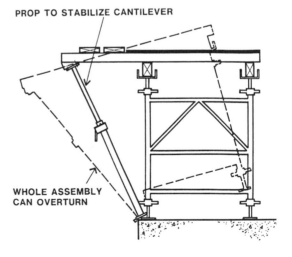

PROP TO STABILIZE CANTILEVER

WHOLE ASSEMBLY CAN OVERTURN

Figure 7.44 - PROPPING TO THE CANTILEVER ENDS

Propping to the Cantilever Forms

The possibility of this type of overturning failure should be regarded as totally unacceptable in formwork construction. One common method that gives greater resistance to overturning is shown in Figure 7.45.

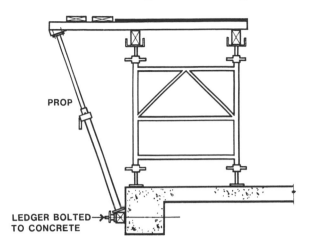

Figure 7.45 - ALTERNATIVE PROPPING PROCEDURE

A timber ledger has been bolted to the face of the concrete edge beam of the lower floor. Props supporting the end of the cantilevering beam are seated on and fixed to this ledger. The top of the prop and all framing members must be fixed. This broadens the base of support and thereby increases the resistance to overturning. However, unless the formwork framing is very heavy, this may still not have sufficient stability to cater for the range of forces that can act on formwork. More effective methods of achieving stability are needed. A method that does this is shown in Figure 7.46.

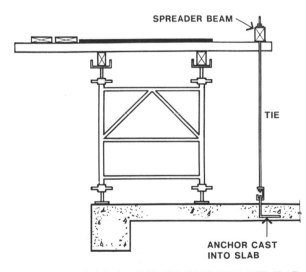

Figure 7.46 - TIE-DOWN ANCHORS CAST INTO THE SLAB

A line of anchors is cast into the slab as hold-down points for the ties. As it is difficult to accurately predict the positions of supports to the edge beam forms, the ties connect to spreader beams which hold down the inner ends of the cantilever beams. Every part of this beam, spreader beam, tie and anchor assembly, and fixings must be adequate to resist the very considerable impact and other loads that formwork can be subject to.

The anchors must be able to reliably resist shock loads. For this reason, expansion type bolt anchors and chemical anchors are not suitable. In such a potentially hazardous situation they are too prone to human error in installation to be reliable. Cast in fittings, like that shown in Figure 7.47, give more reliable anchorage if they are of adequate size.

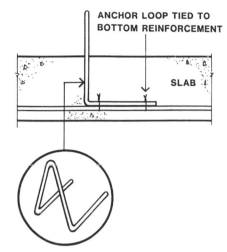

Figure 7.47 - DETAIL OF CAST-IN ANCHOR

After use, these loops can be cut off flush with the concrete surface and the end of the steel bars coated with epoxy to inhibit corrosion. If the building is to have some type of floor coverings or applied finish, this practice is usually acceptable. In other cases an anchor where the top part is removable must be used. Figure 7.48 shows an example.

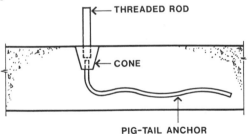

Figure 7.48 - REMOVABLE TIE-DOWN ANCHOR

This anchor uses some wall formwork hardware. (refer to Chapter 4) A pig-tail anchor is bent and embedded alongside the bottom reinforcement of the slab. A He-bolt cone connects to this at the slab surface. A threaded rod, which can be the tie rod, connects into the cone. After the stripping of the beam and soffit forms, the threaded rod and the cone can be removed and the floor patched.

The tie from the anchor up to the spreader beam must also be substantial and adequate for the possible shock loads. For this reason, steel tie wire, even thick wire, is quite unsuitable. Wire has a tendency to fail at twisted wire-to-wire connections. It should be regarded as dangerous if used in this situation. Quality threaded rods, if in good condition and adequate in size, are suitable for these ties.

The spreader beams and cantilever beams must also be adequate in size. If double timbers are to be used for the spreader beams, similar to walers for wall forms, then they should be cleated or bolted together to prevent dislodgement. The reaction from the ties should be spread through large washers.

As an alternative to anchoring the cantilever member at the upper level, they can be installed at the concrete slab level. This is shown in Figure 7.49.

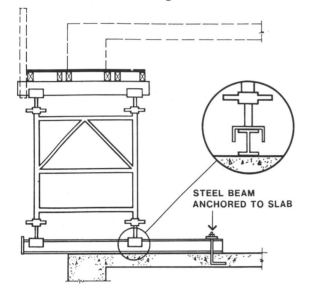

Figure 7.49 - SUPPORTS ON CANTILEVER STEEL BEAMS

Cantilever steel beams, located under the formwork support frames, are each anchored with two 'L' bolt and a top gusset at their inner ends. To ensure that the bases of the support frames will not be dislodged from the top of the cantilevering beams, they are fitted with adjustable 'U' heads inverted over the beams, and the beams with stop-ends. The support frames are located centrally beneath the edge beam, and its formwork can be quite conventional.

If the formwork to the lower floor is still in place when the next level is being formed, it can be used to stabilise the cantilevering platform. The stability of each particular case must be checked. If found suitable it is a most economical way of achieving stability. An example of this is shown in Figure 7.50.

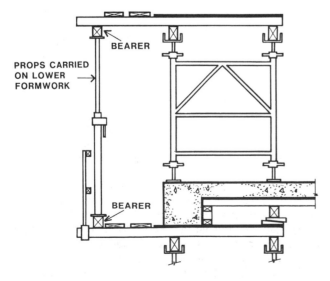

Figure 7.50 - PROPPING OFF THE LOWER FLOOR

The formwork of the lower floor, being concreted, effectively holds the cantilever members in place. Propping can be placed from it up to the canterlevers of the upper level formwork. The props must be securely fixed at their ends to prevent dislodgement. To provide an adequate bearing for the base of the props, a bearer should be installed on the top of the existing cantilever beams.

Upstand Beams

Forms for beams that stand up above the general slab level, require attention to the means of supporting the beam sides and controlling the beam alignment and width. Similar to the step forms shown in Figures 6.82, 6.83 and 6.84, the beam sides can be carried on tie rods and cones from the wall formwork hardware range. Figure 7.51 shows the cross section of an upstand beam that uses that particular technique.

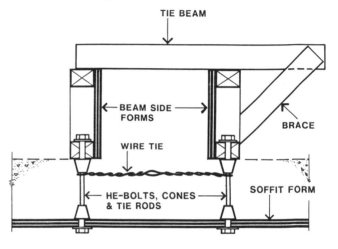

Figure 7.51 - FORMS FOR AN INTERNAL UPSTAND BEAM

The tie rods, cones and bolts give good control over the vertical position of the formwork. If the beam is not very tall, the concrete pressures will be low, and the width can be controlled by wire ties at the base and tie beams at the top. A diagonal brace from the extended tie beam to the form assists in maintaining the squareness of the form.

However, control over the plan alignment of the beam form is not positive. The tie rods are flexible and the whole beam form can drift a short distance off line in either direction. This alignment can be improved by some attention to the construction techniques.

Firstly, the reinforcement cage inside the form can assist if more than usual number of bar chairs are installed between the reinforcement and the beam sides. Secondly, long straight lengths of timber for the top and bottom plates to the forms, and their careful lapping, will ensure that any sideways disturbance at one point will be spread over the whole form.

Finally, care should be taken with the concrete placement. The slab concrete should be placed and vibrated under the beam form first, then the upstand part of the form should be placed and vibrated in short sections to minimize any tendency to misalign the form.

If the upstand beam is near the edge forms, this can be used to fix braces to the top and bottom of the beam form and control alignment. The example shown in Figure 7.52 shows bar ties and walers being used to control the beam width. This technique is suitable for tall beams which have much greater concrete pressures. Where the braces fix to the edge form, a packer has been fixed to ensure trowel access under the brace.

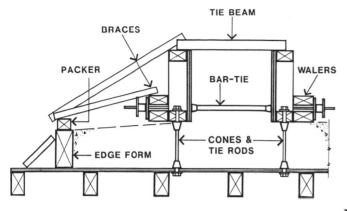

Figure 7.52 - UPSTAND BEAM NEAR THE SLAB EDGE

Where the upstand beam is also an edge beam then the bracing and support problems are simplified. As shown in Figure 7.53, the brace to the outer form cantilevers to support the inner form. Vertical members extend up to complete the framing with top tie beams. Ties are also installed between the tops of the beam sides.

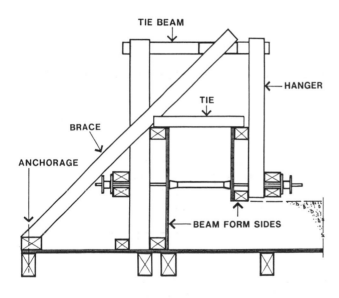

Figure 7.53 - FORMWORK FOR AN UPSTAND EDGE BEAM

In theory, this timber framing can control the width of the beam at the slab level. However, it is likely that unacceptable deflection movement will occur here under the concrete pressure. For this reason it is recomended that a row of ties and walers be installed at this level to give close control over the beam width.

The bracing and form support problems of upstand edge beams are a case which is most suitable for the use of purpose-made steel frames. Considerable economies can result if there are a number of repetitive pours with edge beams of the same size. A suitable steel frame is shown in Figure 7.54. The steel frame can be much stiffer than a timber one, and with a suitably robust frame, horizontal ties are not needed at the bottom of the inner form.

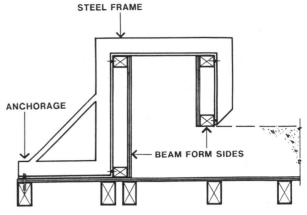

Figure 7.54 - STEEL SUPPORT AND BRACING FRAME

Proprietary Beam Formwork Systems.

The proprietary equipment available for beam framing is generally limited to steel adjustable beam framing devices, often known as beam clamps or adjustable beams. Figure 7.55 shows one of these devices.

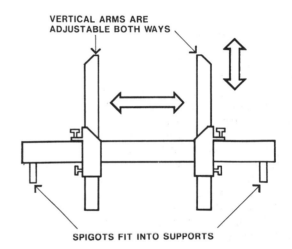

Figure 7.55 - PROPRIETARY BEAM SUPPORT DEVICE

These adjustable beams are intended to be used directly on top of standard framing systems to carry the beam formwork base and side panels. The beam formwork spans between the adjustable beam clamps. The clamps not only carry the vertical loads that act on the beam forms, but also control the beam width and provide resistance to the horizontal concrete pressures.

Figure 7.56 shows an example of a cross section of a beam formed with steel framed plywood faced modular panels for its sides and base. The panels are connected at their adjoining corners with double wedges to external corner angles.

The sets of panels, which are the same length as the distance between the centres of the frames, join to each other over the adjustable beams. Double steel connector wedges are used to hold the one set to the next. Reference should be made to the manufacturer's literature for guidance on their load/span characteristics.

These modular panel products, which are mainly used for wall forms, were described in Chapter 4.

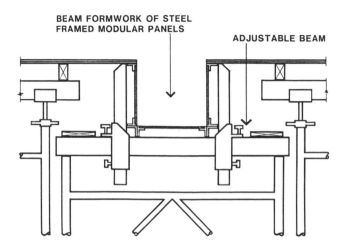

Figure 7.56 - BEAM FORM MADE OF MODULAR PANELS

The adjustable beam clamps can also be used to carry timber and plywood narrow-based beam forms of the type shown in Figure 7.12. A cross section of this case is given in Figure 7.57.

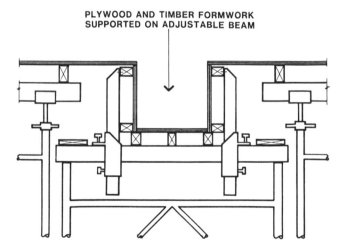

Figure 7.57 - TIMBER FORMS ON ADJUSTABLE BEAMS

Due to the wide spacing between the beam clamps and the resulting large horizontal loads loads acting on the beam sides, the beam side forms usually have to be more robust than those used in the type of forms shown for wide-based beam forms. Those conventional forms are able to have more closely spaced braces which results in less sideways loads on the beam side braces.

In both cases, modular forms and timber forms, there has to be provision for safe access for formworkers to places alongside the beam form. Both details show scaffold planks giving this access area. A discussed previously, the narrow access afforded by scaffold planks is not as convenient, or as conducive to productivity, as a wide plywood surface.

Beam clamps can also be fitted with supplementary props above the legs of the support framing. As noted previously and illustrated in Figure 7.10, these support part of the soffit forms while minimising the bending loads on the beam formwork framing, in this case the beam clamps. An example of this is shown in Figure 7.58.

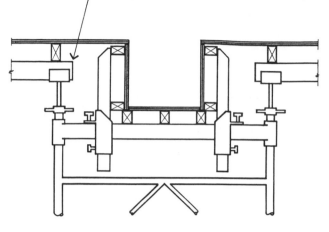

Figure 7.58 - SUPPORT OF BEARERS ONTO BEAM FRAMES

Concrete Encasing Steelwork

Steelwork is concrete encased for two reasons: to fire rate the steelwork and as one of the means of integrating the structural action of the concrete slab and the steel beam. The various types of beam formwork previously described in this chapter can be used to form the concrete encasing of steelwork, but, as shown for slabs in Chapter 6, there can be advantages in suspending all the formwork from the steel frame.

Details were given of the saddle hangers used, and the mandatory requirement to eliminate all overhangs of the saddle hanger legs at the beam edges. This feature, and the means of eliminating it by placing it in a skew position, were illustrated in Figures 6.100 and 6.101.

Figure 7.59 shows the cross section of the encasement of a steel beam and the suspension of quite long span floor centres from the beam form.

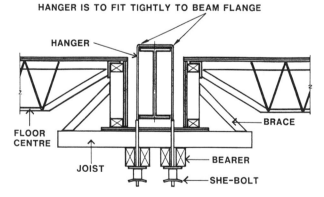

Figure 7.59 - FORMWORK CARRIED ON STEEL BEAMS

This formwork is both stable and symetrically loaded when fully concreted. However, during the progress of the concreting it can be both unstable and dangerous. The worst scenario can occur when the concreting proceeds from one side. This is illustrated in Figure 7.60.

The concrete load, being entirely from one side, causes a large torsional action on the form. At the least, this will distort and misalign the form; at the worst it can cause failure. Two matters must be attended to: the improvement of the torsional resistance of the form, and the adoption of concrete placement techniques that minimise the torsional load.

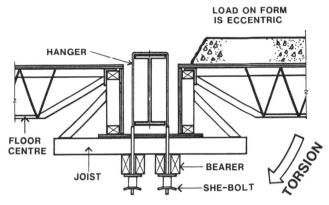

Figure 7.60 - DANGEROUS TORSIONAL LOADING

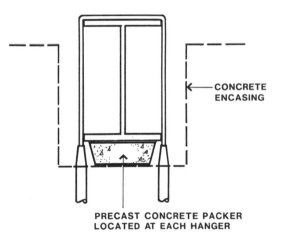

Figure 7.61 - CLAMPING STEEL BEAM AND FORMS TOGETHER

An appreciable part of the torsional resistance of the steel beams can be added to that of the forms by using the She-bolts and saddle hangers to clamp the two together. Figure 7.61 shows one method. Precast concrete packers are placed between the underside of the steel beams and the forms, and the two can then be clamped tightly together. The packers are tapered to improve their retention in place. Further, because the clamping action can crack the packers, it is recomended that they be made of fibre reinforced concrete.

It has been stated previously that the formworker cannot expect to be able to dictate the direction and technique of concrete placement. Nevertheless, if this type of formwork is to be used, the concrete placement must be controlled.

The torsional load will be largely eliminated if the progress of the concrete placement is symetrical about the beam line. Figure 7.62 shows a plan of the 'front' of the pour progressing generally at right angles to the beam. Close control of this will help keep the loading symetrical and minimise torsion. To further enhance the torsional resistance of the assembly, the concrete filling around the steel beam should proceed ahead of the slab pour. However, this should be done in short sections. Normally the encasing concrete is placed down one side of the beam, vibrated it until it appears on the other side, and

then concrete is placed on this second side. This type of placing causes a torsional load in itself, hence the need to do it in short sections.

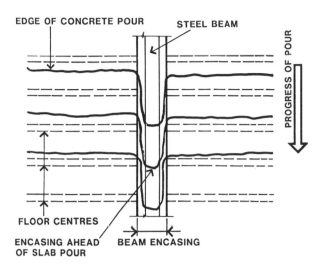

Figure 7.62 - NECESSARY POURING PROCEDURE

MULTIPLE TEE BEAMS

Very early in the history of reinforced concrete it was realised that most of the concrete on the tension side of the slab could be removed with welcome weight benefits. The sections of Figure 7.63 show the principle behind this.

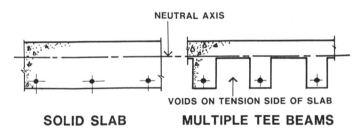

SOLID SLAB   MULTIPLE TEE BEAMS

Figure 7.63 - THE PRINCIPLES OF MULTIPLE TEE BEAMS

The concrete alongside the positive reinforcement in a solid slab contributes very little to the total bending strength of the slab, but when formed, as shown in the right-hand section, it becomes a lighter slab with greater available strength for resisting superimposed loads. These multiple Tee beams provide a method of achieving light deep floor sections that can efficiently span longer distances than solid slabs.

In the early days the voids were formed by building in terra-cotta blocks or similar permanent forms. These methods are still in use and are covered in Chapter 9. Some methods of forming these shapes with removeable forms are given here.

Obviously, multiple Tee beams can be formed by the conventional techniques involving plywood and timber, but the costs are very high. Efficient methods of achieving these shapes depend on the Tee beam forms being re-useable with the minimum of repair. Such forms can be made of metal or plastic. In general, the steel ones are largely self-supporting and the plastic ones require a supporting structure.

Figure 7.64 shows the cross section of forms made of folded steel sheet. With the limitation that the units must be light enough to be able to be man-handled, deeper sections made of thicker steel can span longer distances between supports. Figure 7.65 shows an example of a longitudinal section of the forms.

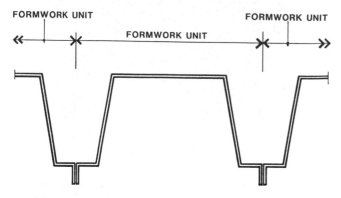

Figure 7.64 - SECTION OF METAL TEE BEAM FORMS

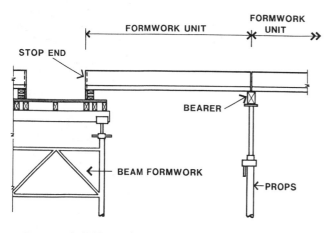

Figure 7.65 - LONGITUDINAL SECTION OF METAL FORMS

At the sides of the beam forms, and the ends of the runs of the units, stop-end are fitted. These must readily separate from the unit as it is stripped. This stripping is usually achieved by flexing the units inwards. The cross section of Figure 7.66 shows a unit fitted with turnbuckles that can pull the sides inwards. Each unit has at least two turnbuckles to enable stripping.

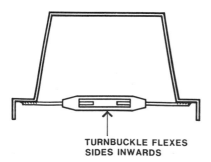

Figure 7.66 - PROVISION FOR STRIPPING METAL FORMS

Units of this type are available in a range of depths and unit lengths, usually all having the same width. As a proprietary product, with prescribed shapes, their use on a structure is effectively dictated when the shape that they

produce is adopted in the project documentation. Some plastic systems have this same feature of modular uniformity.

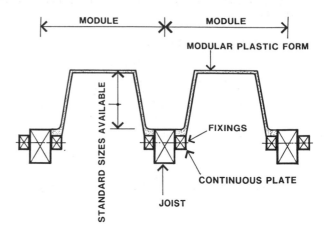

Figure 7.67 - PLASTIC FORMS CARRIED ON TIMBERS

An example of such a modular plastic form and its supporting framing is shown in Figure 7.67. In this case the plastic 'boxes' span between continuous plates on the sides of timber joists. By removing the plates, the plastic form units can stripped out without disturbing the joists that are carrying the concrete ribs of the multiple Tee beam floor.

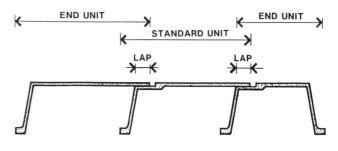

Figure 7.68 - LONG'L SECTION OF PLASTIC FORMS

A longitudinal section (Figure 7.68) shows the lapping of the units together with the special end units. The approximate length of each row of units is generally controlled by the length of the standard units. However, some variation of the laps are possible, and these can accumulate to give a small amount of control over the length of a row of units.

The use of these plastic forms, like the use of the steel forms mentioned previously, dictates the shape of the multiple tee beams. These restrictions on the shape of the concrete, on the one hand, and the source of the formwork, on the other, are often unnacceptable.

As a result there has been a strong emphasis on the development of systems that can provide economic multiple Tee beam formwork for any dimensions. One answer has been found in adapting rigid foam plastic blocks. In both the examples shown, the method starts with the provision of a flat soffit form to support the foam blocks.

The first example involves sheathing the block in a tough, moisture resistant, disposable, cardboard cover. (Figure 7.69) The cardboard covers are taped at the corners but fit loosely over the block. The location of these relatively light blocks is secured with plywood spacer strips fixed to the soffit form. The cardboard cover is easily pierced by the feet of ordinary bar chairs carrying the reinforcement. Only bar chairs with 'saucer' bases should be used.

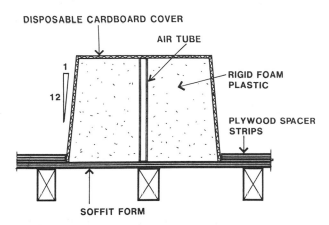

Figure 7.69 - CARDBOARD COVERED FOAM BLOCK

As the cross section in Figure 7.69 shows the block is tapered to enable its withdrawal. Even with this, some force is needed to commence the movement, so a plastic tube is set into the block for air, from a compressed air nozzle, to enter between the foam block and the underside of the cardboard. With the downward movement started, the blocks are easily withdrawn. To keep the blocks to a size that can be readily man-handled, and minimise thermal expansion, their length is usually limited to about two metres. For their next use the blocks are fitted with a new cardboard cover.

The second example is a development of the first. The disposable cardboard cover is replaced with a permanent plastic one made from thin rigid sheet, e.g. ABS, also taped at all corners and joins. Reinforcement bar chairs with 'saucer' bases are also needed here.

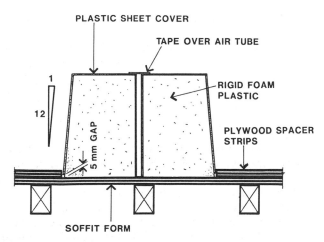

Figure 7.70 - PLASTIC COVERED FOAM BLOCK

Two detail aspects are different to the first example. As the detail in Figure 7.70 shows, the rigid plastic sheet stops 5 mm from the bottom. This is to obviate buckling of the sheet when the foam compresses under load. Secondly, the top of the air tube goes through the plastic sheet, and tape must be placed over it at each use. This is to prevent the tube being blocked with concrete.

Some damage occurs in use, usually during steel fixing. This requires repair before each re-use and limits their economic life to 8 - 10 uses.

Manufacturing Plastic Covered Foam Blocks

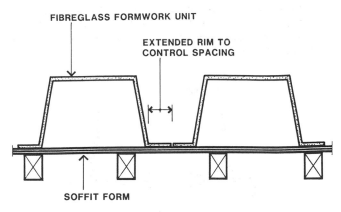

Plastic Covered Foam Block Tee Beam Forms

For a greater number of uses, forms made of fibreglass (GRP) can often be economic. As the section shows (Figure 7.71), control of the location of the forms units can be achieved by extending the rims of the forms to meet at the centre of the concrete ribs. This also provides a useful stiffened edge to the form. The forms are tapered for ease of stripping.

Figure 7.71 - GRP MULTIPLE TEE BEAM FORMWORK

Alternatively, the units can be located as detailed in Figure 7.72. Short clouts are driven into the soffit form at the edge of the unit. Usually 10 mm penetration into the plywood is sufficient. When the soffit is stripped the clouts readily pull out of the plywood and the clouts drop out when the units are stripped.

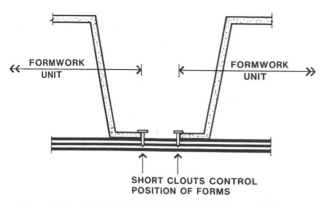

FORMWORK UNIT FORMWORK UNIT

SHORT CLOUTS CONTROL
POSITION OF FORMS

Figure 7.72 - FIXING GRP FORMS TO THE PLYWOOD

Apart from the high cost of the moulds, which have to be spread over the total number of GRP forms, the principal cost in fibreglass forms is usually that of the materials. With deep multiple Tee beams, the concrete pressures place large bending loads on the faces of the units with a resultant rise in thickness, weight and cost. Thinner, lighter units result if the formfaces are stiffened by the addition of ribs made of foam overlaid with fibreglass. These are shown in longitudinal and cross section in Figure 7.73.

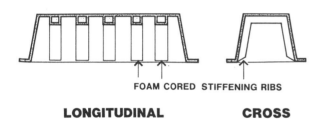

FOAM CORED STIFFENING RIBS

LONGITUDINAL SECTION **CROSS SECTION**

Figure 7.73 - GRP FORMS WITH STIFFENING RIBS

There is normally very little bond between the polyester of the fibreglass and the concrete. Usually, only light applications of release agent are needed. Stripping can usually be started by flexing the unit from the edges. For large units, a hole for a compressed air spike may be needed at the centre.

WAFFLE SLABS:
TWO WAY MULTIPLE TEE BEAMS

Waffle slabs, as the name implies, are simply a grid of multiple tee beams in two directions at right angles. This system is usually adopted by the design engineer when long span floors need to have a striking visual quality. The technique of construction has the advantage of distributing the loads in two directions. As a structural sytem it is very tolerant of high point loads, an advantage for exhibition buildings.

Waffle forms, also known as pans, can be made by a number of methods. Most of the techniques and precautions described for multiple Tee beams are relevant to waffle forms. However, there are very few waffle form products that can be simply supported on a grid of timber beams. Almost all require a flat soffit form as a starting point.

For very large waffle forms, timber and plywood construction is sometimes used. Its expense is only justified if the project specification calls for high quality concrete surfaces, and the large number of uses dictate this type of formwork construction.

Timber and Plywood Waffle Forms

Because of the large size of the plywood and timber waffle forms shown in the photograph, it was anticipated that the initiation of stripping would be difficult to do without defacing the formfaces. To cater for this, an air hole was provided in the middle of the top. It was taped over before concrete placement. To maintain their position, plywood strips were nailed to the soffit form between the pans.

Problems with stripping waffle forms are not confined to timber and plywood ones. The matter must be considered for all types. Because they are generally square units they have proportionately more area of contact with the concrete. To ease stripping, tapers should be greater, at least 1 in 8.

Small waffle pans are often fabricated from sheet metal. In this case a direct downwards force can be an effective stripping method. The inside of the pan is fitted with two connection points for hooks. These are usually located in diagonally opposite corners. After the flat soffit form has been stripped out, the pans can be removed. A weight, with two chains and hooks attached, is hooked onto the connection points inside the pan. The weight is tossed a short distance into the air. The impact effect, as the weight descends and the chains go tight, wrenches the waffle pan out of the concrete.

The Underside of a GRP Waffle Form

GRP Waffle Forms on Flat Soffit Form

The photographs show the shape and installation of GRP waffle forms. The use of this type of waffle form will depend upon having enough forms in each pour, and enough re-uses to justify the production mould, higher and waffle form material costs.

Stripping GRP waffle forms is usually a combination of two actions. For a start, the edge flanges are eased slightly, just enough to break away any grout that is tending to hold them in. This action must never be so boisterous that the edges of the GRP are shattered. After this, compressed air is applied to the central spike hole to force the unit downwards.

GRP waffleforms are compatible with the production of very high quality concrete surfaces, and they can consistently do this over a very large number of uses. The GRP units are very durable. Even if they are damaged they can be readily repaired.

It is common to find large numbers of previously used GRP waffle forms stored in warehouses. If the formworkers can avail themselves of these, then waffle slab formwork costs can be very much lower than expected.

The information given above, on waffle forms, is not intended to be either complete or exhaustive. Many ingenious methods have been tried to find the 'best', 'cheapest' and 'most durable' method of producing this type of concrete construction. It will probably be a continuing quest.

CHAPTER 8: STAIR FORMS

Many formworkers regard stair forms as difficult and complex to construct. This need not be so. Stair forms are merely a special type of sloping soffit formwork. The special aspects are the need to cater for the extra forces that result from the slope and the techniques used to mould the shape of the steps. As will be shown, an important part of the details of the step formwork lies in assisting the achievement of an accurate finish to the concrete surface of the stair treads.

To understand stair formwork it is necessary to appreciate how it carries the loads imposed on it. The forces and the reactions that act on a typical stair soffit form are shown diagramatically in Figure 8.01

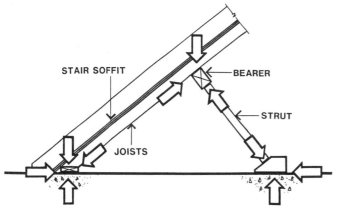

Figure 8.01 - LOADS ON STAIR FORMS

In this case a sloping soffit, anchored at the bottom, is supported part way up the slope on a strutted bearer set at right-angles to the slope. This arrangement results in horizontal force reactions that call for the anchoring of all the formwork supports. Within the form structure, adequate fixing between all members is needed to maintain the load path.

Strutting to the Sloping Soffit Form

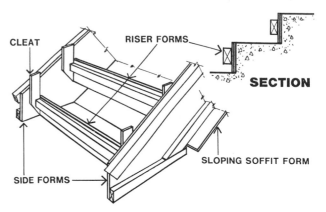

Figure 8.02 - LOWER END OF A STAIR FORM

Good strong side forms are needed to provide rigid and accurate supports for the riser forms. These can be single sections of timber or, as shown in Figure 8.02, plywood strips edged with timber.

It will be noted that no braces to these sides are shown. Only a minimum of braces are needed, and the sole function of these is to hold the side forms plumb. The lateral forces from the fluid concrete pressure, acting on the sides, are catered for by the riser forms, and the fixings to the sloping soffit form. These collectively tie the two sides of the stair form together.

In addition to this tying action, the riser forms must carry the lateral pressure of the fluid concrete that tends to bend them outwards. For the narrow stairs shown, the riser forms span between the side forms. To cater for this sideways bending action, the riser forms must be suitably stiff. They are shown made from a plywood strip reinforced with a timber beam. For wider stairs, a central support and tie is also needed. One example of this is shown in Figure 8.03.

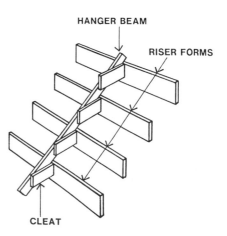

Figure 8.03 - CENTRAL TIE TO RISERS

The hanger beam provides both a tie and a support, and the cleats maintain the line of the riser forms and prevent them twisting. To be an effective tie and thus prevent the sideways deflection of the riser forms, the hanger beam must be anchored against longitudinal movement. Often this can be a strut to a nearby wall, or a fixing to the slab at the lower end.

When stairs are being poured, the concretors walk on the riser forms as they place the concrete and finish off the treads. The central hanger beam is a very effective means of distributing the worker's weight over a number of the riser forms, thereby minimising their deflection under the load.

When close control over the accuracy of line and plumb of deep and wide riser forms is called for, a more efficient method of holding the middle of the riser forms is usually needed. An example of this would be seating to a grandstand.

One method of achieving this is shown in section in Figure 8.04. The hanger and strut connections, from the riser form to the hanger beam, give very positive control over the line and plumb of the risers.

Stair Riser Forms Fixed to Plate on Wall

The problem of accurately finishing the concrete surface to the treads has been mentioned earlier. Being a sloping form, there is a tendency for the concrete to slump down the sloping soffit form and 'waterfall' over the tread forms. As a result, the surface of the treads can be very rough. Often this does not matter in cases where the treads are specified to be 'topped' at a later stage.

Where a quality surface finish is specified for the structural concrete of the treads, then good placing techniques and stiff concrete can help in the control of these inherent problems. Nevertheless, it is hard for the concrete finisher to get an accurate line at the base of the riser form if the riser form is a simple rectangular shape. The two problems that can arise are shown in the sections of Figure 8.06.

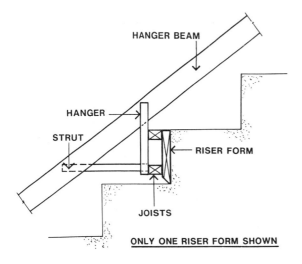

Figure 8.04 - DEEP RISER FORMS

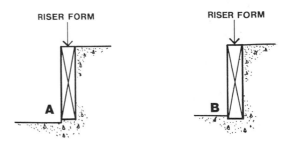

Figure 8.06 - SECTIONS OF RISER FORMS

Where the stairs are to be constructed against a face of a wall, the riser forms must be supported on the wall. Figure 8.05 shows a longitudinal section of the stair forms. The ends of the riser forms are suspended, each by a hanger cleat, from a timber plate. The timber plate is usually shot-set to the wall.

If the concrete surface of the tread is lower than the underside of the riser form, there will be a step in the tread. (Figure 8.06 'A') If it is higher, a groove results. (Figure 8.06 'B') The formworker can assist the concrete finisher by adopting better details for the bottom of the riser form. The simplest answer is shown in Figure 8.07. This detail was included in Figure 8.04 above.

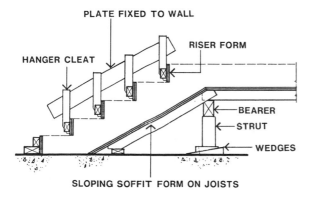

Figure 8.05 - RISER CONNECTION TO WALL

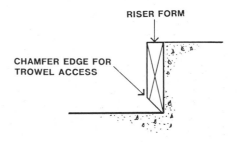

Figure 8.07 - IMPROVED STAIR RISER

The chamfering of the edge, for almost its full width, permits trowelling of the tread closer to the riser face. The potential for error is greatly reduced. However, even when the lower edge of the riser is chamfered, it still has some width, rarely less than 3mm. This can still show on the finished concrete surface. Sheet metal can be used as an aid to hiding this mark.

Two examples of this use of sheet metal are shown in section in the details shown in Figure 8.08.

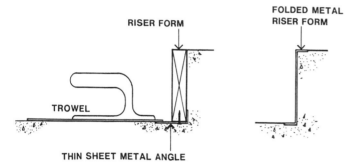

Figure 8.08 - METAL EDGED RISERS

The left-hand section of a tread shows a simple rectangular riser form fitted with a thin sheet metal angle. The trowel can easily be lined up with the thin metal edge to give an accurate surface. The one defect that cannot be avoided is the minor setdown that is fixed by the thickness of the metal of the angle. The right-hand detail gives a development of this with the whole riser form folded from sheet metal. This riser form detail is used in a permanent formwork system for stairs shown in Chapter 9, Figure 9.38.

Not all riser shapes are simple flat vertical surfaces. An example of the riser form for a different shape is given is the section of Figure 8.09.

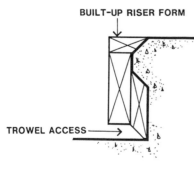

Figure 8.09 - BUILT UP RISER FORM

The shape is shown built-up from a number of special shaped timber sections. This particular shape of riser could also be efficiently formed from folded sheet metal.

PRECAST STAIRWAYS

In many buildings, and in particular in multi-storey buildings, there are often a great number of identical stair flights. In these cases, the advantages of precasting the stairs merit consideration. This does not only mean precasting in an off-site factory, it also includes site precasting. The pouring of the small concrete quantities involved is usually done at the same time as the pouring of in-situ elements such as floors, columns and walls. Figure 8.10 shows the sectional elevation of a typical precast stair installation.

Figure 8.10 - PRECAST CONCRETE STAIRS

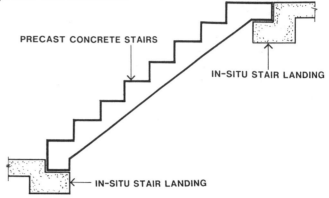

Precast Stairs Ready for Installation

Precast stairs can be readily installed onto the newly cast floor structure, usually the day after the floor is poured. It only requires the stripping of the edge forms; the slab soffit and beam forms need not be disturbed. Access to the stair for site personel is then immmediately available. In contrast, with in-situ stair construction, this traffic is impeded by the supports to the stair soffit forms.

Because these stair units can be cast upside-down or on their edge, very good accuracy and a high quality surface finish can be consistently achieved. Further, even though the forms are expensive, their repetitive use, and the resultant faster construction cycle, can yield significant overall economies.

Form for Casting Stairs Upside-Down

An example of formwork for a precast stair unit which is to be cast on its edge is shown section in Figure 8.11 and plan view in Figure 8.12.

Edge casting has the advantage of taking less space than upside-down casting; an important consideration on a crowded building site. Further, the vertical form faces to the top and bottom of the stairs can be stripped early for re-use on another flat surface. In temperate to warm climates, this stripping time can be 18 hours. With upside-down casting it is normal to wait at least four days before lifting the precast stair unit off the forms.

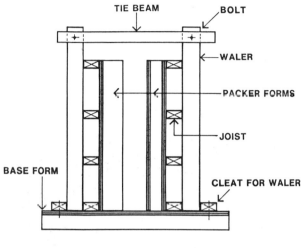

SECTION

Figure 8.11 - SECTION OF PRECAST STAIR FORM

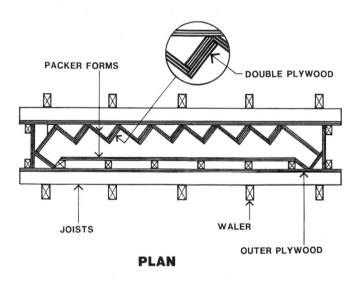

PLAN

Figure 8.12 - PLAN OF PRECAST STAIR FORM

CHAPTER 9: PERMANENT FORMWORK

Not all formwork is removed when the concrete no longer requires support. There can be advantages in leaving it in place, even if these are only in the elimination of stripping costs. This formwork is generally known as **PERMANENT** formwork but sometimes as **LOST** or **SACRIFICIAL** formwork. As will be seen, in some cases these alternate labels can sometimes be more appropriate.

At their simplest, permanent forms, like other formwork, are only a temporary support to control the shape of the fluid concrete until it gains strength. However, they can also be devised to contribute to the strength of the reinforced concrete structure, or provide a high quality surface for it. Precast concrete units can be used as permanent formwork for both these purposes.

In the first and simple case, the only restrictions on the formworker's selection of the permanent form material may be weight and durability limitations. In the other two cases, the important decisions on material selection and details of construction are made by the designer of the permanent structure. The formworker plays only the lesser role of devising any required temporary support and alignment framing.

The arrangement of this chapter on permanent forms closely follows the order of the preceding part of the book dealing generally with removable formwork. The materials for permanent formwork are discussed first. Details of the construction of the several types then follows in the same order as the earlier chapters.

MATERIALS FOR PERMANENT FORMS

Many materials are used in permanent formwork. The most common are timber, galvanized sheet steel, concrete, glass reinforced cement, cardboard and plastic. The applicability of each of these materials will depend on the requirements of durability, strength and appearance for each case.

A number of general parameters can be given for the materials for permanent forms:

1. They must be strong enough to carry the pressures that the fluid concrete can exert.
2. At a minimum, this strength must endure until the concrete is self-supporting. For soffit forms this must be at least six or seven days and may in some cases need to be more. With vertical formfaces ten to twelve hours may be sufficient.
3. Any breakdown of the form material, including corrosion, must not be incompatible with the structural action or durability of permanent reinforced concrete structure. In some cases the breakdown is necessary, for example, with the void formers used to create a gap between suspended slabs and expansive soils, this breakdown of strength is essential.

1. Timber

Because of its characteristics, timber is not one of the most suitable materials for permanent forms. Most timbers exhibit significant swelling with increases in their moisture content, and great care must be taken in their use. If this expansion is restrained, the resultant reaction forces may cause damage to the permanent structure. An example of the detrimental effect of timber swelling is illustrated in the section shown in Figure 9.01.

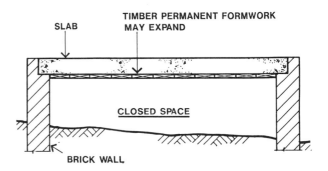

Figure 9.01 - INCORRECT USE OF TIMBER FORMS

Timber boards are shown used in permanent formwork to a slab. As the under-slab space is enclosed, the moisture in there often builds up and the boards expand. There have been cases where the resultant force on the walls has been enough to push them outwards. A technique to avoid this is shown in Figure 9.02.

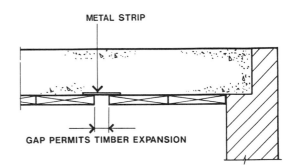

Figure 9.02 - EXPANSION GAP BETWEEN BOARDS

Here, expansion gaps have been constructed using metal strips. These relieve any swelling of the timber. If the permanent soffit form used is plywood, there will be very little lateral expansion but the thickness of the plywood may markedly increase with swelling.

Expansion is not the only problem that comes with moisture increase. Excess moisture can also cause timber to rot. The resultant odours may be unacceptable.

In summary, timber should only be used in permanent formwork with care and caution.

2. Galvanized Sheet Steel.

This material is strong, generally durable, and readily adaptable to permanent formwork. Not only are the 'corrugated' roofing products suitable for serving a purely support function, but, there is a range of proprietary shapes available, where the ribs of the sheets also act as reinforcement to the concrete.

However, excessive moisture over a period of time may promote corrosion and a breakdown of the galvanised sheets. This feature may make galvanized steel unsuitable for a simple fluid concrete support function. With the form/reinforcement systems this is not a matter for the formworker to decide. These are specified in each case by the project designer.

3. Concrete.

Both insitu concrete and precast concrete units are used for permanent formwork. In the simplest case, insitu concrete can be used as a blinding slab on the ground. An example of this is a suspended structure built over soft ground that is expected to settle over a long period. The blinding slab can provide an accurate surface that will give the concrete the correct shape, and also provide a firm base for bar chairs to the reinforcement.

For permanent formwork, precast concrete offers a wide range of uses. Its durability and general compatability with the insitu concrete makes it ideal in many respects. Precast units can provide fully or partially self-spanning soffit forms that can contribute to the strength of the permanent structure. They can provide high quality concrete surfaces with accurately detailed shapes that aid waterproofing and weathering patterns and that can interface with other elements such as curtain walls on multi-storey buildings.

In both these cases, the task of the on-site formworker is usually limited to providing framing to support the precast units and control their alignment. Where they have high quality surface finishes, effective protection against abrasion, impact and staining may be needed. This protection includes catering for abrasion that can occur between the concrete face and the support framing.

The design of these units, either for structural or architectural requirements, is not usually the task of the formworker. As they are a significant part of the permanent structure their detailed design must be included in the project documentation. The support, alignment and protection requirements are also normally specified in this documentation. These may be given as very detailed instructions or merely specified in performance terms.

The instructions often include requirements on storage, handling and erection. These are often critical. To minimise the weight of the permanent forms, the designer takes advantage of the ability of the precasting process to produce thin sections. Thin elements made of high strength concrete can be quite brittle. They are prone to damage from impact, poor handling and storage, and rough erection practices.

The detailed design of precast formwork units is beyond the scope of this book. As noted previously, this is the job of the project designer. The information given in this chapter will be limited to general descriptions, with the emphasis being on the formworker's tasks.

4. GRC - Glass Reinforced Cement.

GRC products are essentially a rich cement/sand mortar reinforced with alkali resistant glass fibres. Like the similar process of manufacturing fibreglass (glass reinforced plastic GRP) they are made by simultaneously spraying the mortar and fibres onto a single faced mould.

The mortar provides a good level of compressive strength and the fibres the tensile strength. The result is a relatively light, durable form unit, which is quite compatible with the insitu concrete. In the range of shapes that can be produced, the process is more versatile than precast concrete. Indeed, the shape is only limited by the requirements of spraying onto and stripping from a single faced mould.

Like precast concrete forms, GRC forms require careful attention to their support, alignment and protection. They are almost always used for their high quality external surface finish. Further, the surface may be prepainted. Special consideration is needed for protection of these surfaces.

5. High Density Fibre Reinforced Sheet.

Cement based, fibre reinforced, rigid sheet materials are widely used in the buiding industry. For permanent formwork, high density (compressed) sheets, reinforced with cellulose fibre, are available in thicknesses from 18 to 40 mm.

To optimise their load-to-span characteristics, the fibres are placed dominantly in one direction. This 'grain' is clearly marked on the sheets, and for safety and maximum load carrying capacity it must run across the span. Good load/span characteristics are available. For example, data from one manufacturer shows that a 18 mm thick sheet will support a 200 mm slab over a 450 mm span and a 40 mm sheet will carry a 300 mm slab over 1100 mm.

The material, being cement based, is compatible with the insitu concrete and is manufactured with a high quality external finish. It is very durable and does not usually require any surface protection.

6. Rigid Foam Plastic.

Where a light-weight durable void former is needed, rigid foam plastic is often suitable. There are, however, two important characteristics to be catered for.

Firstly, being very light, its flotation must be catered for. When the vibration has fluidized the concrete, the uplift can be up to 26 kN per cubic metre of foam void form. In contrast its dead load can be as low as 1.6 kN per cubic metre.

The second problem is its vulnerability to damage from immersion (poker) vibrators. If impact with vibrators occurs, pieces of the plastic foam can break off. This is unnacceptable to the project designer. The problem can usually be avoided by wrapping the piece of rigid foam in tough flexible sheet plastic. Joints and edges should be sealed with strong packaging tape.

7. Cardboard.

All of the previous examples of materials for permanent forms are quite durable, in some cases as duarable as the permanent structure. With cardboard void formers, it is expected that its durability need only extend to the time needed for the concrete to develop sufficient strength to be self supporting. This time rarely exceeds 6 to 10 days.

For void formers under suspended slabs, it is often essential that the cardboard form should collapse so that it cannot transmit any future uplift forces to the concrete structure. To achieve this short finite structural life, the cardboard form product is given a waterproofing treatment that looses effectiveness in the continuous presence of moisture from the ground below.

In other cases, such as tubular void forms within the slab, the collapse of the cardboard is not essential to the permanent structure.

EXAMPLES OF PERMANENT FORMWORK

It is not possible to cover every case of permanent formwork. The following examples are only intended to be a representative range of the available techniques.

1. Ground Forms

The most common use of permanent formwork for concrete work on the ground, is in the creation of voids between the soffit of the structure and the ground. These voids spaces are usually needed to isolate the structure from any uplift forces. The forces occur as a result of the ground getting wetter and then swelling. Figure 9.03 shows the simple example of permanent void formwork for footing beams which are carried on deep piers.

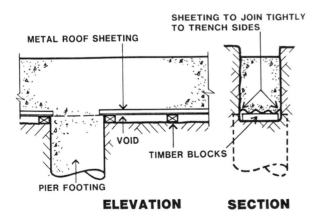

Figure 9.03 - VOID FORMWORK FOR GROUND BEAMS

Corrugated sheetmetal, usually one of the standard galvanized roofing profiles, is used for the formface and it spans over a series of timber blocks. The metal sheeting must be fitted tightly to the sides of the trench to prevent any concrete or grout filling the void space below the form. The technique is not confined to beam work. It can be applied to larger areas such as slab soffits.

Soil expansion uplift forces that can act on the beam soffit are not completely eliminated. A small proportion of the uplift can still be transmitted through the timber blocks. These may take a long time to rot away. If this situation is unacceptable, a total void can be created by using cardboard forms that will collapse after a short time. Two types of these void formers are shown in Figure 9.04.

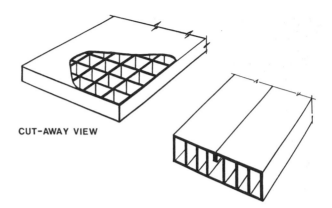

CUT-AWAY VIEW

Figure 9.04 - EXAMPLES OF CARDBOARD VOID FORMS

The forms can be of 'egg-crate' construction, as shown in the left-hand detail, or be cut and folded into a cellular form from sheet cardboard.

As noted previously, they are intended to gradually lose crushing strength in moist conditions. However, the large amount of water that is present in the fluid concrete when it is placed, will cause the cardboard to collapse almost immediately. Accordingly, the forms should be protected from this excess water by installing a waterproof membrane, such as thin sheet plastic, over the forms. Under no circumstances should there be any membrane that will inhibit the eventual rise of ground moisture up into the cardboard. Figure 9.05 shows the section of a typical case.

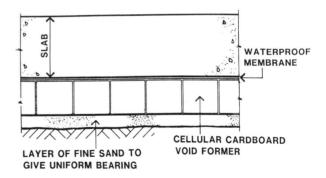

Figure 9.05 - CARDBOARD VOID FORMS UNDER A SLAB

Underneath the void forms a layer of sand has been installed. This is to give the forms a uniform bearing surface and maximise their load carrying capacity. If the cardboard boxes were to be seated on uneven ground, the resultant twisting action and concentrations of loads could cause their collapse during the concrete pour.

2. Wall Forms.

The most common material used for permanent formwork for walls is precast concrete. It can be used for single faced wall forms or as one or both faces of double faced formwork. One example of the use of precast concrete panels for single faced forms is shown in Figure 9.06.

Mass concrete dams are usually too wide for the efficient use of double faced formwork. While the vertical face can be readily formed by the repetitive use of a single faced form, as described in Chapter 4, the outer curved face is more complex. Precast concrete as permanent formwork can simplify matters by providing a formwork system that does not need access to the outer face.

INNER FORMS NOT SHOWN

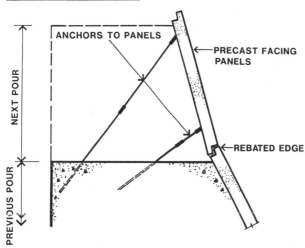

Figure 9.06 - PRECAST SINGLE FACED FORMWORK

At each lift, the slope of the panel can be set to follow the intended curve, and its base located with a rebated joint. The precast panel normally has anchor points cast into it. These are used to connect to adjustable ties, which have been built into the previous pour. The panel is designed to resist the concrete pressures and span in two directions between the tie points. This method of anchoring to the inner side of the panel is adopted to avoid the need to provide formworker access to the outer face of the panels.

The alternative is to provide holes in the panels for the installation of removeable ties. Figure 9.07 shows the section of a double faced formwork arrangement with through ties.

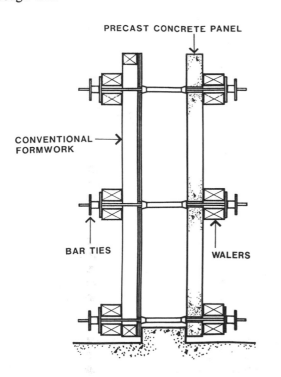

Figure 9.07 - PRECAST WITH CONVENTIONAL FORMS

One face uses permanent precast concrete form panels, and the other conventional wall forms. The structural action on the precast panel is limited, in this case, to spanning vertically. The necessary strength between the ties, in the horizontal direction, is provided by the walers.

While this makes the panels lighter and cheaper, it also means that the tie rod pattern is readily visible, and the tie rod holes holes must be patched or plugged. A further cost is the provision of formworker access to the outer faces. For these cost and aesthetic reasons, the concealed fixings are more commonly adopted.

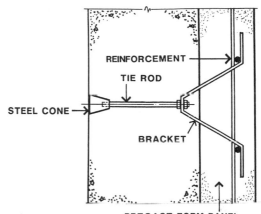

Figure 9.08 - CONCEALED TIE ROD FIXING - 1

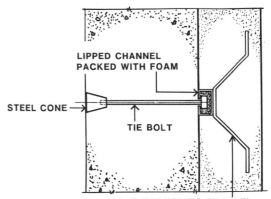

Figure 9.09 - CONCEALED TIE ROD FIXING - 2

Figures 9.08 and 9.09 show cross sections of two methods of making concealed fixings to precast permanent formwork panels. In the first, the tie rods are double nutted onto the cast-in brackets. The location of the formwork wall ties are controlled by the positions of the brackets. In the second, the provision of continuous lipped channels permits tie bolt installation anywhere along the channels.

The cast-in lipped channels have a further advantage for the project designer. In many cases it is desireable to ensure that, in the future, the precast panel can move relative to the in-situ concrete. This movement could be thermal, concrete shrinkage or structural action. To enable this to occur, the channels are placed to suit the expected movement, the inner face of the precast concrete is debonded before installation, and the channel is filled with foam plastic to prevent it becoming clogged with concrete.

The penetration of grout would tend to prevent the movement, and cause undesirable secondary stresses in the concrete.

3. Column Forms.

The simplest example of permanent column forms is the use of pipes to form circular columns. Two types in current use are fibre reinforced and precast concrete pipes. These are standard industrial products made for hydraulic work.

The fibre reinforced pipes, being manufactured for water supply work, are usually able to carry the fluid concrete pressures. They are normally available in lengths up to four metres and usually only require bracing near the top. For columns taller than four metres, pipe jointing is needed. However, a very neat joint is difficult.

They are suitable for most normal concrete work, but, because of the nature of their external finish, they are not usually satisfactory when high quality concrete surfaces are specified.

Butt jointed reinforced concrete pipes do not have these external finish and jointing limitations. Figure 9.10 shows cross sections through the joints between pipes known as 'Internal flush' and 'External flush' joints. This pipe nomeclature relates to the provision of a groove which is to be filled with jointing material to achieve a flush joint. For an 'Internal flush" joint, the groove is internal. It is for the project designer to specify which of these joints is to be used.

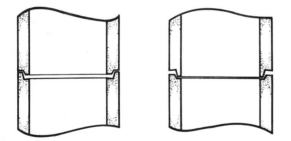

INTERNAL FLUSH JOINT   EXTERNAL FLUSH JOINT

Figure 9.10 - PRECAST CONCRETE PIPE JOINTS

The pipes are normally available in 1.2 and 2.4 metre lengths and various classes of strength. The pipes can be readily cut to length with standard concrete cutting equipment. To adequately cater for concrete pressures it is recommended that the highest pressure class be used.

Concrete pipes usually have a good finish marred only by the stencilled name of the manufacturer. However, the products can usually be obtained without this label.

For columns using only one length of pipe, propping near to the top is all that is needed. If more than one length of pipe is used, the joint must be clamped for alignment. This alignment involves two matters, straightness of the whole column and alignment of one pipe circle over the other. Figure 9.11 shows one suitable method of achieving this.

Four timbers are clamped onto the pipe sections with column clamps. The timbers must be adequate, in size and length, to provide bending strength across the joint as well as alignment between the pipes.

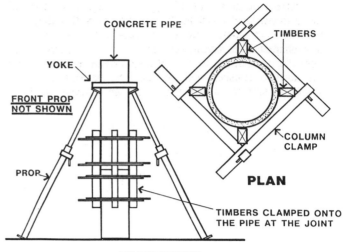

Figure 9.11 - ALIGNMENT OF CONCRETE PIPES

Columns rectangular in section, as well as special shapes, can be formed with precast concrete. The permanent forms can completely enclose the shape as shown in Figure 9.12 or be used in conjunction with re-useable forms as shown in Figure 9.13.

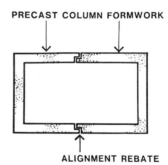

Figure 9.12 - PRECAST FORMS TO ALL COLUMN FACES

Precast Forms to all Column Faces

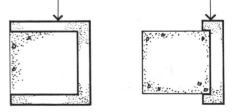

Figure 9.13 - PARTIAL FORMING WITH PRECAST FORMS

Permanent forms of this type are adopted because of the efficient and consistently accurate way that they can provide high quality external surfaces. As noted earlier in this chapter, the formworker must not only support and align the permanent forms but also provide protection for the surface finishes and details. The permanent forms shown in section in Figure 9.14 are an example of this.

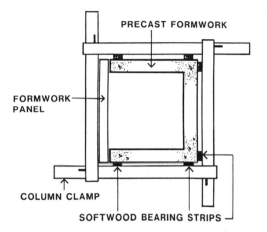

Figure 9.14 - CLAMPING PRECAST CONCRETE FORMS

To contain the open shaped precast forms against the concrete pressures standard column clamps are used in conjunction with a conventional formwork panel for the open side. To preserve the texture of the concrete surface, the high crushing stresses between the clamps and the surface are reduced by placing softwood bearing strips at the interface. The use of timber walers and tie rods, also caters for this need for large bearing areas on the precast concrete faces. (Figure 9.15)

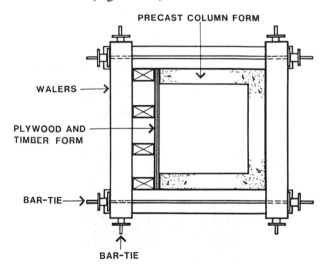

Figure 9.15 - CLAMPING WITH WALERS AND TIE RODS

The task of support and alignment can be as simple as bracing conventional re-useable column forms. However, with precast column forms at the edge of a structure, it is often more complex. The tolerances for alignment of a column with the one below are often very small. To cater for these accuracy requirements, the support frame, usually made of steel, should be fitted with appropriate two-direction adjustment mechanisms. A diagramatic illustration of an example of this is shown in Figure 9.16.

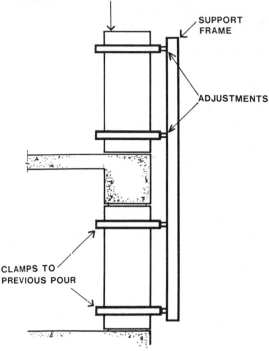

Figure 9.16 - ALIGNMENT OF EXTERNAL COLUMNS

The support frame clamps onto the previously cast column. The column clamps for the next column are connected to this frame through two levels of adjustable connections. Through these connections the column form can be moved sideways in two directions by using both levels of adjustable connections in the same directions. Also, by using the upper and lower adjustments, the form can be accurately plumbed.

4. Soffit Forms.

We can apply two broad categories to permanent soffit forms: those that only provide support for the fluid concrete to achieve the desired shape, and those that perform the additional function of contributing to the strength of the permanent structure.

In the first category, for the construction of bridges, precast units can used be to form the insitu slabs that span between the main precast concrete girders.

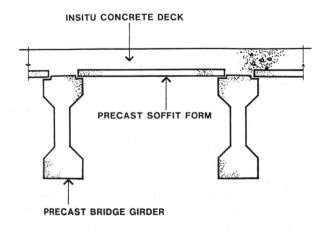

Figure 9.17 - PRECAST SOFFIT FORMS TO A BRIDGE

Figure 9.17 shows the cross section of permanent precast concrete soffit forms for a bridge. Permanent formwork has attractions in this case as the stripping of conventional formwork under the bridge deck can be both expensive and hazardous.

The precast soffit form slabs are normally 50 mm or more thick and their weight, as part of the permanent structure, causes an increase in bridge construction costs. The use of flat high density fibre reinforced sheet results in a reduction in this dead load. A further reduction in load can be obtained by using GRC (glass reinforced cement) and foam plastic to achieve the weight reduction advantages of multiple Tee beams. (Figure 9.18)

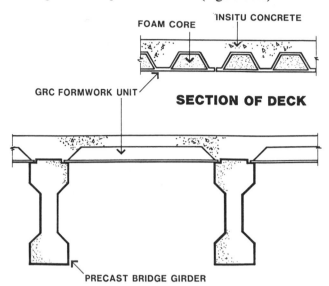

Figure 9.18 - GRC AND FOAM SOFFIT FORM UNITS

As Figure 9.18 shows, the GRC forms have a flat soffit, foam plastic cores and a shaped GRC top face. This shape creates light but strong formwork to support the concrete and construction loads. The resultant multiple-Tee-beam shape reduces the concrete quantities. It also reduces the bridge loads without loss of strength of the bridge structure. However, the formwork system is expensive.

A number of other examples of permanent forms for multiple Tee beams, generally used in buildings, also lie in this first category. Using a flat soffit form, hollow terra-cotta or concrete blocks are used to form the voids between the beams. (Figure 9.19)

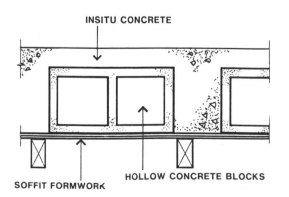

Figure 9.19 - HOLLOW BLOCK VOID FORM UNITS

Another example of void formwork, given in Figure 9.20, takes advantage of the adaptability and economy of fibrous plaster. Fibrous plaster, plaster-of-paris reinforced with plant fibres, can be moulded to virtually any shape and the product is quite compatible with concrete. Further, they are sufficiently durable for permanent formwork. Experience has shown that they can resist up to one weeks heavy rain before showing signs of softening. If consistent rain is anticipated, a water-proofing agent can be incorporated in the material.

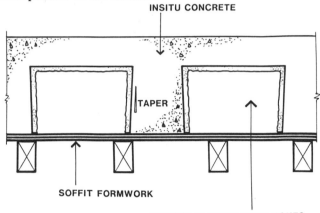

Figure 9.20 - SECTION OF FIBROUS PLASTER FORMS

The four sided, open bottomed fibrous plaster boxes are placed on a flat soffit form. The long sides have a small vertical taper to ensure that the forms do not drop out in the future. The spaces between the rows of boxes form the multiple Tee beams. The space between the ends of the boxes forms the cross ribs.

The low cost and simplicity of the process of producing these boxes enables the production of a range of sizes and shapes to suit each particular case. For example, the boxes can be shaped to produce tapered concrete beams. This gives an increase in their shear strength where it is needed. Figure 9.21 gives an example of a part plan of the formwork boxes in position.

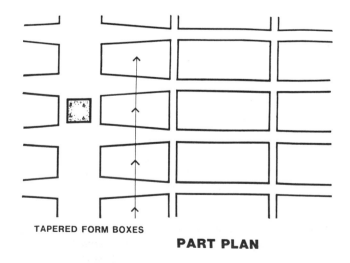

Figure 9.21 - PART PLAN OF PLASTER FORM BOXES

The plaster blocks are simple to produce. Their moulds can be made from pieces of plywood hinged at their junctions on the base, for removal of the block. The sides are connected with simple hooks at the corners. (Figure 9.22)

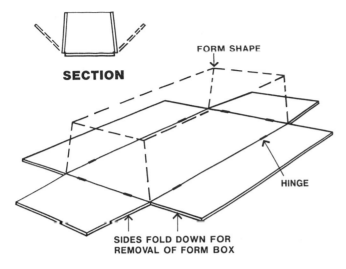

Figure 9.22 - MOULD FOR PLASTER FORM BOXES

After mixing, the plaster of paris and fibre (usually hemp) is dumped into the mould and shaped by hand up the sides. The mixture hardens quickly, and further amounts of the mix are placed to extend the sides up to the top. The material thickness varies from 10 mm to 20 mm. For larger boxes the edges usually need some strengthening. As shown in the cross section of Figure 9.23, this is achieved with a timber frame moulded into the plaster work.

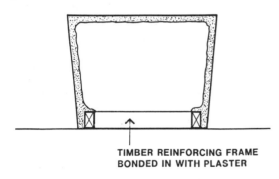

Figure 9.23 - STRENGTHENING FRAME TO EDGES

After several hours the plaster boxes can be removed from the mould after lowering the hinged sides.

The boxes are heavy, usually two formworkers are needed to place them. Their weight assists in holding them in position, and the chairing of the reinforcement off the boxes adds to this effect. If further retention is needed, nails can be driven into the formface alongside the blocks, or spacer timbers can be fixed to the soffit forms, inside the boxes.

These blocks are cheap and simple to make, but they have one major disadvantage. Their appearance is very rough and they are only suitable where a ceiling is to be fitted.

An example of permanent formwork for multiple Tee beams, that produces a good quality soffit, is the use of carboard tubes to form voids in the slab. A civil engineering use for these is found in large box culverts.

These need a smooth flush soffit to ensure that any water borne debris does not catch on the structure. A cross section of the concrete structure is given in Figure 9.24.

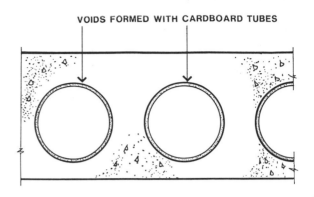

Figure 9.24 - CARDBOARD TUBE VOID FORMS

Cardboard tubes, similar to those used for column forms, are fitted with metal end caps to form cylindrical void shapes. Their outsides are waterproofed with either plastic or wax coating. As a light hollow void form, they are subject to large flotation forces and must be anchored at the correct level. Figure 9.25 shows the section of the tubular void forms in place.

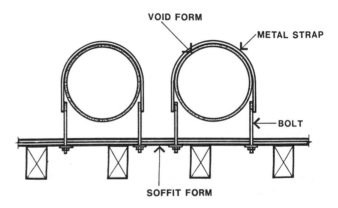

Figure 9.25 - ANCHORAGE OF VOID FORM TUBES

The anchoring straps are needed to tie the forms down. The spacing of these straps, which are never less than 1.2m apart, is given in the manufacturer's literature. For larger tubes, the spacing can be as close as 600 mm. Wires must not be used for the straps; the large flotation forces will cause them to cut into the cardboard. The manufacturer's literature usually advises on the minimum width of the straps.

In addition to the tie downs, supports are needed under the tubes. These can be bar chairs directly under the tubes or bar chairs carrying additional reinforcing bars at right angles to the tubes.

As noted above, the second broad category of permanent soffit forms are those that contribute to the strength of the concrete structure.

A very simple example is the fully self-supporting precast concrete floor unit, which is topped with concrete after installation. Figure 9.26 shows a typical section of these inverted trough units. The topping integrates the structural action of adjacent floor units through the concrete filled shear keys on their sides.

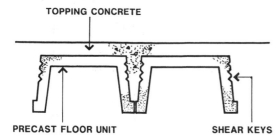

Figure 9.26 - PRECAST CONCRETE FLOOR UNITS

Normally, the framing of the edge forms for these units is supported from the units. The section given in Figure 9.27 shows the framing suspended on bolts through the decks of the units. In other cases the bolts may connect through the vertical webs of the units. The location of these bolts is usually determined by the project designer, whose requirements in this respect must be completely satisfied.

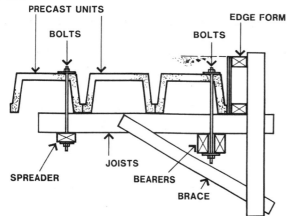

Figure 9.27 - EDGE FORMS TO PRECAST TROUGH UNITS

A cross section of an edge form construction for hollow precast floor units is given in Figure 9.28. The edge forms are held on by He-bolts that connect through rod connectors to hook bolts. With the shafts of the He-bolts embedded deeply in the concrete as shown, they will need to be heavily greased or fitted with plastic tubes to enable removal.

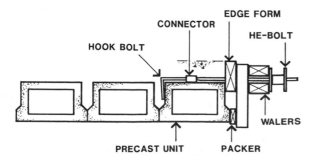

Figure 9.28 - EDGE FORMS TO HOLLOW PRECAST UNITS

The widely used range of types of metal deck units are another example of the type of permanent formwork that contributes to the permanent structure. The concrete bonds

to the units which are then able to act as reinforcement to the concrete. Figure 9.29 shows the cross sections of three of the types of galvanized sheet metal floor units.

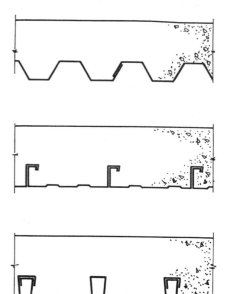

Figure 9.29 - THREE EXAMPLES OF METAL FLOOR UNITS

To effectively act as reinforcement, the units must be correctly installed at their seating on support walls and junctions to beams (Figure 9.30). The usual minimum is 50 mm bearing in these cases. When seated on steel beams, it is normal practice to connect them to the beams by puddle welding through the flat tray of the unit. Information on the frequency and size of these welds should be obtained from the manufacturer of the floor units.

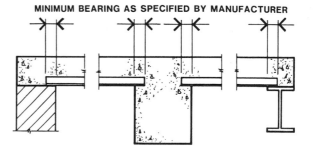

Figure 9.30 - END BEARING REQUIREMENTS

These floor units have limited ability to carry the load of the fluid concrete and the construction activity. The manufacturer's literature gives data on the distance that the units can safely span, relative to the thickness of the slab to be poured.

Other than for the case of short spans, one or more lines of intermediate supports are needed. These intermediate supports are usually timber bearers on props or frames. Figure 9.31 shows examples of cross sections through both thes types of supports. The conditions for their removal are given by the manufacturer. These normally call for a concrete strength of at least 75% of the target 28 day strength.

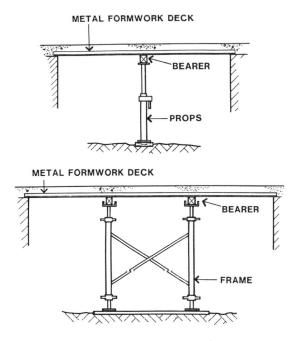

Figure 9.31 - INTERMEDIATE SUPPORT SYSTEMS

Slab edges can be formed by conventional means as shown in the section of Figure 9.32 or with the proprietary sheet metal edge form section shown in Figure 9.33.

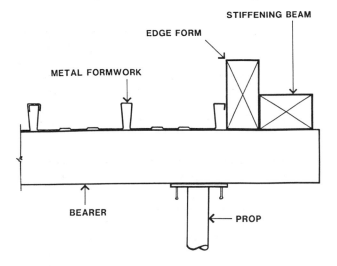

Figure 9.32 - SECTION OF TIMBER EDGE FORMS

As the edge form does not have a soffit form surface to which it can be fixed, its sideways deflection must be controlled by the installation of a stiffening beam. As the detail shows, the bearers are extended past the edge forms to provide support to them.

Figure 9.33 shows two sectional views of the use of a proprietary sheet metal edge forms, one parallel to the ribs and the other at right-angles. Where the edge form is parallel to the ribs, it has been found that the outer rib of the floor unit interferes with pouring a solid concrete edge. For this reason, it is often removed.

The sheet metal edge forms have a stiffened top edge. In both cases, to maintain alignment, the top of this top edge needs to be tied to the deck at regular intervals. Galvanized metal straps are screwed or pop rivetted to the ribs as shown.

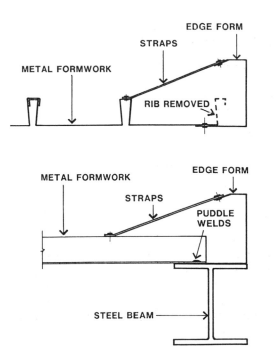

Figure 9.33 - PROPRIETARY SHEET METAL EDGE FORMS

Penetrations through the slab are often needed for ducts and services. Provided appropriate extra trimming reinforcement is placed around the penetration, a number of ribs can be cut to make the hole. Additional bearers are required to support the ends of these ribs and, as shown in Figure 9.34, the penetration edge formwork can be fixed to them.

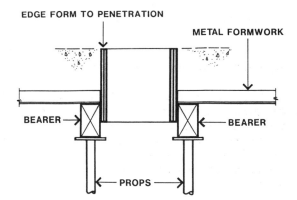

Figure 9.34 - EXTRA SUPPORTS AT PENETRATIONS

5. Beam Forms

Unlike column and soffit work, which have a range of permanent formwork products available, this work for beams is generally confined to decorative facings manufactured in precast concrete. Figure 9.35 shows the cross section of the edge beam of the insitu concrete structure with a precast permanent formwork facing.

The support structure and associated removable formwork is shown in Figure 9.36. The framing and its bracing must cater for the fluid concrete pressures on both the permanent forms and the conventional forms. The bracing must also prevent any sideways movement. Misalignment from any cause could not be repaired.

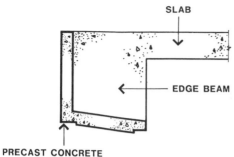

Figure 9.35 - PRECAST FORM FACING TO BEAM

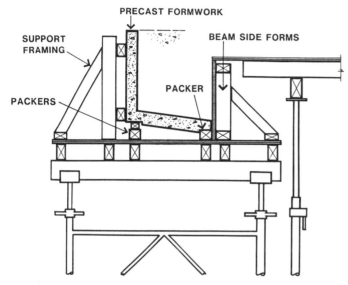

Figure 9.36 - SUPPORT OF PRECAST BEAM FACING

The arrangement of the support of the precast forms must also allow for the cleaning away of any grout loss at their joints. It is common practice to hose this grout away during the pour. Direct lines of access for the hose jet are needed. Note that there are two levels of packers at the outer edge of the precast units. The upper level packers have gaps coinciding with the joints in the precast units. The gaps give access for the hose jet.

6. Stair Forms.

The sloping soffits of stair slabs can be efficiently formed with the proprietary galvanised sheet metal floor units described previously. A longitudinal section is shown in Figure 9.37.

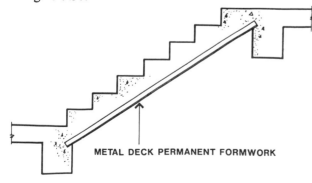

Figure 9.37 - PERMANENT METAL FORMS TO SOFFIT

This type of formwork is also incorporated into a totally prefabricated sheet metal stair form system which is complete with sheet metal side forms and riser forms. Figure 9.38 illustrates this type of form.

These permanent forms are not confined to standard 'risers' and 'goings', they are available tailor-made to suit the particular project. The only framing that they need is end bearing and intermediate supports, to suit the requirements of the metal soffit forms. The end bearing arrangements must also prevent any sliding of the form.

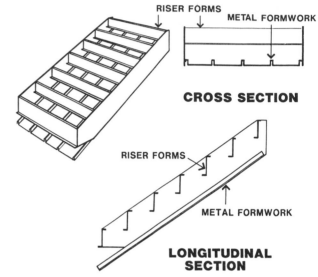

Figure 9.38 PREFABRICATED METAL STAIR FORM

Sheet Metal Stair Form in Place

As the photograph indicates, but is not shown on the diagram, the permanent metal stair form is delivered to site with reinforcing mesh already in place. The stair form shown is being installed between the faces of two concrete walls. Its sideways alignment is being controlled by wedges between the side faces and the walls.

****

This chapter has been limited to an overview of this fast developing area of formwork technology. Permanent formwork systems can offer economy and speed of construction, as well as the advantages of quality finishes and making a contribution to the strength of the structure.

BIBLIOGRAPHY

British Standard BS 5268: Part 2: 1989
Structural Use of Timber – Part 2 – Code of Practice for permissible stress design, materials and workmanship.

British Standard BS 5975: 1982
Code of Practice for Falsework.

Council of Forest Industries of British Columbia
Canadian COFI Exterior Plywood for Concrete Formwork – Engineering information.

INDEX

A

Abrasion
  coil ties 42
  plywood 17, 20, 87, 104
Access panels for walls 70
Access platforms 64, 76, 105
Accuracy 16, 26
  columns 86
Adjustment joints
  column gang forms 107
Aluminium 19
Aluminium beams 19, 77, 145
  as walers 78
  quick strip systems 149

B

Backpropping 143
Beam forms 152
  base of beam form 154
    eccentric loading 155
    plywood deflection 155
  collapse 153
  deep beams 158, 159
    with ties 159
  edge beams 160
    multi-storey 162
      anchorage 163
      on steel beams 164
      propped forms 162–164
    restricted space 160
  encasing steelwork 166
  floor centres 156
  grout loss 158
  hangers 166, 167
  horizontal forces 152
  hung off steel beams 166
  loading 152
  load path 153
  load path redirection 154, 166
  load reduction - strutting 157
  permanent - precast 185
  pressures 152
  principles of stripping - sequence 154
  proprietary
    panels 166
    support beams 165
  separate to soffit 157
  sides 156
    braces 155
    joist seating 157
    stripping of sides 158
  stripping 154
  torsional failure 167
  unstable 153
  upstand beams 165
  wide beams 156
Bearers 116
  eccentricity 116

  sizes 117
Bracing 48, 102, 123, 125, 127, 129
  column forms 91, 97
  to concrete column 127
  wall forms 48

C

Cardboard forms
  columns 97
  permanent 177, 178, 183
Care and maintenance 14
Circular column forms 97
  from sheet plastic 98
Climbing wall form 37, 77
Coil tie 42
Collapse 113, 145, 153
Column forms 86, 180
  access platform 105
    to edge column 106
  accuracy requirements 86
  arris 90
  bracing 91, 96, 97, 102
  circular 97
    bracing 97
    steel 99
    stripping cardboard 98
    timber 98
    yoke for 97
  clamp 87
  conventional 88
  deflection 93
  edge connection 94
  erection 90
  gang forms 106
    adjustments 107
  general details 99
  grout loss 101
  high quality
    arris 90
  hinged 87, 94
    eccentric hinge 95
    three hinged 96
    two hinged 95
  hoisting 103
  horizontal walers 91
  integral with soffit form 103
  internal ties 96
  kickerless construction 100
  kickers 100
  perimeter stripping 92
  plywood spans 88, 99
    two way 89
  pressures 86
    from dropping concrete 86
    proprietary 87
  rectangular 87
    large 96
  special shapes 109
  special steel clamps 92

starter bar positions 99
strapping 92
three hinged 96
tolerance at top 101
top of column 101
two hinged 95
two part 93
    steel 94, 99
Compatibility of materials 16
Concrete loads 5
Concrete mounding 7
Concrete pressure
    confined spaces 6
    factors controlling 6
Construction joints 13, 14
Creep of foundations 5
Curved wall forms 78
    cover strip 79
    in straight sections 82
    large radius 79
    proprietary system 80, 82
    small radius 81
    steel tube walers 80

D

Deflection 40, 136, 155
    column form 93
    formface 41, 50
    progressive 4
    wall framing 40
Differential support movement 5
Dowelled joint 27
Durability 16, 176

E

Eccentric footings - beam forms 31
Eccentricity 10, 146, 148, 155
Economy 1, 12, 16, 144, 174
Edge beams 160
    in restricted space 160
    multi-storey 162
Edge forms
    on ground 26
        cantilevered 29
        dowelled 27
        hard ground 28
        keyed 27
        rebated 30
        with top step 29
        with waterstop 28
    on soffit forms 134
Environmental loads 10

F

Failure 11, 16, 42, 112, 113, 118, 120, 128, 133,
    139, 140, 145, 147, 148, 155, 162, 167, 176
Failure mode 9, 11, 16, 37, 102, 113
Fibre reinforced sheet 117, 182
Fixings 23, 46
    nails 23

plywood 49
properties and types 23, 24
screws 24
wall form framing 50
Floor centres 145, 156
    end support 146
    failure 147, 148
    increasing load capacity 147
    minor set downs 148
    stripping 149
Footing forms 30
    eccentric 31
    isolated rectangular 30
    pedestal forms 30
    strip 31
Formface
    GRC 22
    GRP 21, 22
    materials 19
    particle board 22
    plywood 19–21
    solid timber 21
    steel 21
Frame systems – see Soffit forms
Framing materials 18

G

Gang forming of columns 106
    adjustments 107
    sequence of use 108
Glass reinforced cement (GRC) 22
Glass reinforced plastic (GRP) 21
Groove forms
    pilasters 68
    tapered on wall forms 58, 59
Ground forms 26
    permanent 178
    void forms 178
Grout loss 41, 158
Guy ropes 4, 132

H

He bolts 41, 42
Hinged column forms 94
Hoisting
    column forms 103
    fittings 65
    wall forms 64, 76
Horizontal waler
    lapping of 40
    wall forms 38
Hydration staining 17

I

Impact 8, 16, 37, 86, 102, 112
    loads 113
    resistance 16
Information 1, 14
Instability 9, 162

J

Jacks, screwjacks 40, 73, 122, 127
    details 129
Joists 115–116
    buckling 18
    lapping 115
    lateral restraint 18
    sizes 115
    sizing 15, 16

K

Keyed edge forms 27
Kickerless construction
    columns 99
    walls 53
Kickers 32
    columns 99
    function 32
    types 33
    walls 53
    with groove 33

L

'L' beams 152
Learning curve 13
Load action 10
Loads 5
    concrete 1
    construction 8
    environmental 10
    materials 7, 18
    wall forms 36

M

Make-up strips 141
Material
    accuracy 16
    evaluation 18
    stiffness 16
    strength 16
    weight 16
Materials
    aluminium 19
    compatibility 16
    damaged 17
    durability 16
    formface 19
    framing 18
    impact resistance 16
    plywood 9
    rigid foam plastic 22
    steel 19
Materials and components 16
Minor set downs 138, 147
Movement of supports 5
Multi-storey buildings - loads 8
Multi-storey frames 11, 132

Multiple Tee beams 182
    cardboard and foam block 169
    fibrous plaster 182, 183
    GRP 169
    metal forms 168
    plastic covered block 169
    plastic forms 168
    principles 167
    tube void forms 183
    two way 170
    void forms 182

O

Overturning 4, 132
    guy ropes 4

P

Penetration forms
    soffit forms 137
    wall forms 61
Permanent formwork 176
    beam forms 185
    columns 180
    failure 176
    fibre reinforced sheet 177
    fibrous plaster 182, 183
    ground 178
    materials 176
    metal deck units 184
    metal stair forms 186
    multiple Tee beams 182
        void forms 167, 182
    precast concrete
        beam facing 186
        concealed fixings 179
        floor units 184
        single faced wall forms 179
        stairs 174
        wall forms 178
    stairs 186
    tube void forms 183
    void forms 167, 178, 182
Pilaster piers 66
    deep 67
    groove forms for 68
    separate construction 67
    shallow 66
Plastic forms
    circular columns 97
    multiple Tee beams 168
Plywood 19, 49
    columns - two way span 89
    corner of column forms 89
    curving radii 78
    face veneer 20
    facings 20
    fixings 49
    GRP facings 21, 22
    in column forms 88
    lapping of sheets 114
    selection 19, 114

soffit forms 114
types 19–20
Pre-ageing 21
Pressure 6
factors controlling 6
Progressive collapse 12, 113
Progressive errors 59
Proprietary equipment
column forms 87
components 24
wall forms 64
components for large units 73
curved 80
large
components 73
details 74, 77
plywood faced 72
steel 70
steel soldiers 70, 73
Props 123, 150 - see Soffit forms

Q

Quality 1, 16
Quick strip systems 149

R

Release agents 22, 23
chemical 23
mould cream 23
neat oil 23
neat oil with surfactant 23
water soluble emulsion 23
wax emulsion 23
Reshoring 143
Runners - see Bearers

S

Safety 1, 2, 16, 18, 26, 36, 43, 45, 86, 91, 104, 112, 115,
122, 123, 144, 153, 162, 166
Self weight 5
She bolts 43
Sheet metal forms
circular columns 97
rectangular forms 94
Sidesway 4
Single faced wall forms 83
anchorage 46, 83
corner tying 85
overturning 83
sloping 85
tied back 85
Sliding 3
Soffit forms 112
aluminium beams 145
backpropping 143
backpropping strips (bands) 143
base plate rotation 122
bearers 16
fixing to props 124

lapping 116
bracing to permanent structure 123
cantilevering 141
causes of collapse 113
construction details 134
construction joints 135
construction philosophy 113
conventional 114
eccentricity of bearers 116
edge forms 134
floor centres 145
end support 146
failure 146, 147, 148
increasing load capacity 147
loading eccentricity 146
minor set downs 147
footings and foundations 117
on improved ground 121
frame systems 127
bracing 129
connectors 129
eccentricities 130
end rotation 131
erection braces 127
fittings 128
load application 130
frames and props together 8, 133
impact 113
joists 115
lapping 115
uniformity of size 115
junctions to walls 134
loads 112
make-up band 141
minor set downs 138, 147
modular systems 132
erection 133
on the permanent structure 118
on steel beams 139
on uneven surfaces 120
penetrations 137
prop systems 123
base tripod 124
bracing 125
erection procedures 125
quick strip systems 149
erection procedures 150
reshoring 143
slab penetrations 137
sloping formwork 138
failure at the bases 139
spreader under props 120
stages of loading 112
stiffness of sole plates 119, 120
stripping 142
support systems 122
frames 127
props 123
workmanship standards 124
telescopic prop fittings 124
timber framing sizes 115, 117
tolerance gap 142
tolerance problem 141
top step 136

Soffit forms *contd*
    undisturbed supports 142
Snap ties 44
Spandrel wall formwork 54, 55
Special formwork systems 13
Special shaped columns 110
Stability 3, 9
    overturning 4, 162
    guy ropes 4, 132
    sidesway 4
    sliding 3
    uplift 4
Stages of construction 2–3
Stages of loading 3, 8
Stair forms 172
    central tie beam 172
    deep riser forms 173
    loading 172
    metal edged riser forms 174
    metal soffit 186
    precast concrete 174
    prefabricated metal form 186
    riser forms 173
    strutting to soffit 172
Steel 19
Stiffness 3, 16
Stock pile of materials 8
Stop-ends to walls 50
    deflection at 50
        packer studs for 51
    details of construction 51
    proprietary mesh 52
    waterstop 52
Storage racks for wall forms 14, 64
Strength 3, 16
Strip footings 31
    step forms 31
Stripping 142, 144, 149, 154
    circular column 98
    column forms 91
    floor centres 149
    wall forms 37
Structural requirements 3
Support movement 5
    rotation of baseplates 122
Surface finish 1, 15
Suspended ground forms 26, 30, 31, 33, 34

T

'Tee' beams 152
Telescopic props 123
    fittings 124
    fixing to bearer 124
    tripod for 124
    types 123
Through ties 43
Tie bolts 41
Tie rods 41
Timber
    manufactured sections 19
    sizing 15, 16, 46
    solid sections 18

Tolerances 1, 40, 86, 101, 141
Tube props 123
Tunnel form loading 5

U

Undisturbed supports 142
Unstable 9
Uplift 4, 29
Upstand beam 165

V

Vertical waler wall forms 79

W

Waffle slabs 170
Walers 41
Wall forms 36
    access panels 70
    access platforms 64
    bracing 48, 76
        bracing frames 48
    cast-in fittings 61
    climbing 37, 40, 77
    construction details 49
    construction joints 50
    corners 55, 57
    double faced 38
    erection procedures 46–48
    fabrication and erection 46
    framing
        action 39
        fixings 74
        members 38
    groove forms 58
        haunches 69
        pilasters 68
    haunch (ledge) 69
    height of pour 37
    hoisting 36, 64, 76
    kicker 53
    kickerless 53
    lapping of walers 40
    large proprietary components 73
    load paths 38, 39
    loading 36
    penetrations 61
    pilasters 66
    proprietary systems 70
        plywood faced panels 72
        steel panels 70
    recess forms 60
    single faced forms 36, 83
    stepped face 68
    stop-ends 50
    stripping 37
    supporting the base 37
    surface features 58
    tapered walls 69
    Tee junction of walls 57

thickness variations 68
top of pour 63
type selection 40
vertical waler 39
wind load 36
Wall spacers 43
Wall ties 41
   coil ties 42
   He bolts 41

installation 45
She bolts 43
snap ties 44
through ties 43
wire ties 43
Waterstops 28, 52
Weight of materials 16
Windows in wall forms 70
Wind loads 10